Chapter 1: Introduction to Oscill(

- Basic Principles ... 10
- Types of Oscilloscopes ... 10
- The Importance of Digital Oscilloscopes ... 11

... 11

Chapter 2: Evolution of Digital Oscilloscopes ... 12

- Evolution from Analog to Digital ... 12
- Key Milestones ... 12
- Major Contributors ... 13
- Closing Thoughts ... 13

... 13

Chapter 3: Understanding Oscilloscope Specifications ... 15

- Bandwidth ... 15
- Sampling Rate ... 15
- Memory Depth ... 15
- Other Important Features ... 16
- Tips for Selecting the Right Oscilloscope ... 16
- Wrapping Up ... 16

... 17

Chapter 4: Choosing the Right Oscilloscope ... 18

- Factors to Consider ... 18
  - Bandwidth ... 18
  - Sample Rate ... 18
  - Channels ... 18
  - Memory Depth ... 18
- Price Range ... 19
  - Budget Friendly ... 19
  - Mid-Range ... 19
  - High-End ... 19
- Recommended Brands ... 20
  - Tektronix ... 20
  - Keysight Technologies ... 20
  - Rigol ... 20
  - Teledyne LeCroy ... 20

Siglent... 21
Final Thoughts.. 21
.... 21

Chapter 5: Initial Setup and Calibration.. 22
Power Supply... 22
Grounding.. 22
Probe Connections.. 22
Other Initial Setup Steps... 23
In Conclusion... 23
.... 23

Chapter 6: Display Controls and Menus - Your Guide to Navigating the World of Oscilloscopes.. 24
Display controls, control knobs, and menu options.................................. 24
The Display Control Panel.. 24
Control Knobs and Buttons.. 24
The Menu Options.. 24
Understanding the Hierarchical Menus... 25
Navigating the Menus with Soft Keys and Knobs................................... 25
Take Your Time and Experiment.. 26
In Conclusion... 26
.... 26

Chapter 7: Voltage, Frequency, and Time Measurements: An Introduction to Basic Waveform Analysis Techniques.. 27
Voltage Measurements: Understanding the World of Oscilloscopes..... 27
Frequency Measurements: The Music of the Oscilloscope World......... 27
Time Measurements: Capturing the Essence of Signals......................... 28
Basic Waveform Analysis Techniques: The Art of Reading and Interpreting Waveforms.. 28
Conclusion... 28
.... 29

Chapter 8: Saving and Recalling Waveforms.. 30
How to Save and Recall Waveforms.. 30
Tips for Organizing Saved Waveforms... 30
Conclusion... 31
.... 31

Chapter 9: Understanding and Mastering Triggering and Capture Modes... 33
Different Trigger Types and How to Use Them for Specific Applications... 33
Edge Triggering... 33
Pulse Triggering... 33
Video Triggering... 33
Pattern Triggering... 34
Advanced Triggering Techniques... 34
Tips for Effective Triggering... 35
Mastering Triggering and Capture Modes for Accurate Measurements... 35
... 35

Chapter 10: Advanced Measurements and Their Role in Troubleshooting... 36
RMS (Root Mean Square) Measurements... 36
Peak-to-Peak Measurements... 36
Other Advanced Measurements... 37
Using Advanced Measurements for Troubleshooting... 37
Conclusion... 37
... 38

Chapter 11: Probes and Probe Compensation... 39
Types of Probes... 39
    Passive Probes... 39
    Active Probes... 39
    Current Probes... 39
    Differential Probes... 39
Probe Specifications... 39
    Bandwidth... 40
    Rise Time... 40
    Input Impedance... 40
    Attenuation Ratio... 40
Tips for Probe Compensation... 41
    Check the Ground Lead... 41
    Use the Correct Attenuation Setting... 41
    Compensate for Probe's Capacitive Loading... 41
    Minimize Noise and Interference... 41
    Refer to the Oscilloscope Manual... 41

In Conclusion..........41

..........42

Chapter 12: Troubleshooting with Oscilloscopes..........43
Common Troubleshooting Techniques using Oscilloscopes..........43
Finding Ground Loops..........43
Signal Distortions..........43
Takeaways..........44

..........44

Chapter 13: Advanced Features and Applications..........45
FFT Analysis..........45
Serial Protocol Decoding..........45
Other Advanced Features and Applications..........46
Conclusion..........46

..........46

Chapter 14: Saving and Exporting Data from Oscilloscopes..........48
The Importance of Saving and Exporting Data..........48
Options for Saving and Exporting Data..........48
CSV Files..........48
Waveform Image Files..........48
Other File Formats..........49
Tips for Efficient Saving and Exporting..........49
In Conclusion..........49

..........50

Chapter 15: Best Practices for Using Remote Control Interfaces and Automation in Digital Oscilloscopes..........51
The Power of Remote Control Interfaces..........51
Convenient and Efficient Automation..........51
Using Remote Control Interfaces and Automation Together..........52
Streamlining Your Workflow..........52
Best Practices for Using Remote Control Interfaces and Automation..........52
The Future of Remote Control Interfaces and Automation in Oscilloscopes..........53
In Conclusion..........53

Chapter 16: Tips for Maintaining and Calibrating Your Oscilloscope for Accurate Measurements..........54
Tips for Regular Maintenance..........54

- 1. Read the User Manual......54
- 2. Keep It Clean......54
- 3. Check the Connections......54
- 4. Inspect for Damage......54
- 5. Perform Firmware Updates......55

Tips for Calibration......55
- 1. Understand Calibration Terms......55
- 2. Use Calibration Signals......55
- 3. Check the Vertical and Horizontal Scales......55
- 4. Perform Regular Calibration Checks......55

Tips for Troubleshooting Calibration Issues......56
- 1. Inspect the Probes......56
- 2. Check for Loose Connections......56
- 3. Perform a Factory Reset......56
- 4. Send for Professional Calibration......56

Conclusion......56

......57

Chapter 17: Troubleshooting Common Oscilloscope Issues......58

Tips for Solving Noise Issues......58
Tips for Solving Display Issues......59
Tips for Solving Power Issues......59
Tips for Solving Triggering Issues......60
Conclusion......60

......60

Chapter 18: Understanding Trigger Jitter and Noise......62

Sources of Trigger Jitter and Noise......62
External Sources of Trigger Jitter and Noise......62
Internal Sources of Trigger Jitter and Noise......63
Techniques for Minimizing the Impact of Trigger Jitter and Noise......63
- 1. Proper Grounding and Shielding......63
- 2. Use Differential Probes......64
- 3. Adjust Trigger Levels and Slopes......64
- 4. Use High-quality Cables and Connectors......64
- 5. Keep Your Oscilloscope Clean and Well-maintained......64

.... 64

Chapter 19: Further Troubleshooting with Oscilloscopes 65

Tips for Optimizing Your Oscilloscope Usage 65

Choose the Right Probe 65

Properly Compensate Your Probe 65

Use the Correct Settings and Probes for Probing High-Frequency Signals 65

Utilize the Waveform Math Functions 66

Understand and Use Triggering Effectively 66

Take Advantage of Advanced Applications 66

Regularly Calibrate Your Oscilloscope 66

Stay Up-to-Date with Latest Technologies and Trends 67

In Conclusion 67

.... 67

Chapter 20: Common Applications of Oscilloscopes 68

Overview of Applications in Various Fields 68

Electronics 68

Telecommunications 68

Medicine 69

In Conclusion 69

.... 69

Chapter 21: Understanding Waveform Shapes and Patterns in Digital Oscilloscopes 71

The Basics of Understanding Waveform Shapes and Patterns 71

Identifying Specific Components and Parameters 71

Using Your Knowledge of Waveforms to Diagnose Issues 72

Conclusion 72

Chapter 22: Tips for Accurate Measurements 73

Understanding Sampling Rates and Averaging 73

Set Your Trigger Correctly 73

Use Correct Probe Settings 73

Minimize Noise and Interference 74

Adjust Time and Voltage Scales 74

Calibrate Your Oscilloscope 74

Conclusion 74

.... 74

Chapter 23: Reducing Noise and Interference in Oscilloscope Measurements 76
Strategies for Reducing Noise and Interference 76
Proper Grounding Techniques 76
Shielding and Filtering Techniques 77
Signal Routing and Probing Techniques 77
Ground Loops and Differential Probing 77
Conclusion 78
.... 78

Chapter 24: Troubleshooting Digital Circuits 79
Techniques for Troubleshooting Digital Circuits 79
Identifying Faulty Logic Levels 79
Timing Issues in Digital Circuits 79
Other Tips for Troubleshooting Digital Circuits 80
In Conclusion 80
.... 81

Chapter 25: Advanced Measurements with Mixed-Signal Capability 82
Overview of Mixed-Signal Measurement Capabilities 82
How to Set Up and Interpret Results 82
.... 83

Chapter 26: Power Analysis with Oscilloscopes 84
Using Oscilloscopes for Power Measurements 84
AC and DC Power Analysis 84
Power Supply Troubleshooting 84
Conclusion 85
.... 85

Chapter 27: Advanced Applications of Oscilloscopes 86
Comparing and Contrasting the Functionality of Oscilloscopes and Spectrum Analyzers for Frequency Domain Analysis 86
Understanding the Basics of Frequency Domain Analysis 86
The Functionality of Oscilloscopes for Frequency Domain Analysis 86
The Functionality of Spectrum Analyzers for Frequency Domain Analysis 87
The Advantages and Disadvantages of Using Oscilloscopes and Spectrum Analyzers 87
In Conclusion 87
.... 88

Chapter 28: Debugging, Decoding & Analysis..89
Advanced Features for Efficient Debugging..89
Bus Decoding Capabilities..89
Real-time Serial Protocol Analysis..90
Enhanced Measurement Capabilities...90
Conclusion..90

....91

Chapter 29: Advanced Applications of Oscilloscopes...................................92
Techniques for reducing noise and interference in power measurements..........92
Using current probes..92
Differential probes...92
Proper grounding techniques..93
Proper shielding techniques..93
Isolation techniques...94
Keeping the signal path short...94
In Conclusion..94

....95

Chapter 30: Tips for Probing High-Frequency Signals with Precision and Accuracy....96
Tips for Probing High-Frequency Signals without Affecting Signal Quality............96
    Use Active Probes..96
    Use Short Ground Leads...96
    Minimize Wiring and Connections..97
    Pay Attention to the Probe Tip..97
    Use the Appropriate Attenuation Setting..97
    Use Differential Probes for Differential Signals..................................97
    Keep the Probes Close to the Device Under Test..............................98
Conclusion..98

....98

Chapter 31: Techniques for Measuring Signals with Fast Rise Times...............99
Introduction..99
The Importance of Accurate Measurements..99
Choosing the Right Oscilloscope..99
Using a Low-Inductance Probe...100
Adjusting the Attenuation Factor..100

Using a High-Speed Trigger 100
Enabling the Averaging Feature 100
Adjusting the Oscilloscope's Ground Level 101
Using Math Functions for Signal Averaging 101
Conclusion 101

.. 101

Chapter 32: Advanced Troubleshooting Techniques with Waveform Math Functions. 102

Application of Waveform Math Functions to Analyze Complex Signals and Troubleshoot Issues 102
Understanding Waveform Math Functions 102
The Types of Waveform Math Functions and Their Applications 102
    FFT (Fast Fourier Transform) 102
    Differentiation and Integration 103
    Advanced Triggering and Mask Testing 103
Real-World Examples and Case Studies 103
Best Practices for Using Waveform Math Functions 104
In Conclusion 104

.. 104

Chapter 33: Techniques for Troubleshooting Analog Circuits: Identifying Signal Distortions and Component Failures 106

Understanding Signal Distortions 106
Using Waveform Analysis to Identify Distortions 107
Identifying Component Failures 107
Best Practices for Troubleshooting Analog Circuits 107
Conclusion 108

.. 108

Chapter 34: Advanced Applications of Oscilloscopes 109

Case Studies in Automotive Testing 109
Examples of Medical Device Testing 109
Advanced Triggering Options for Specific Measurement Scenarios 110
Pulse Width Triggering 110
Video Triggering 110
Delay Triggering 111
Runt Triggering 111

Logic Triggering.. 111
Combining Triggering Options.. 112
Conclusion.. 112

Chapter 35: Advanced Techniques for Characterizing Signal Integrity..................... 114
What is Signal Integrity?... 114
Eye Diagrams for Characterizing Signal Integrity.. 114
Jitter Analysis for Characterizing Signal Integrity....................................... 115
Tools for Advanced Signal Integrity Analysis... 115
Conclusion.. 115
.... 116

Chapter 36: Advanced Bus Analysis for Troubleshooting and Debugging............... 117
Advanced Bus Analysis Techniques... 117
I2C Bus Analysis... 117
SPI Bus Analysis... 117
CAN Bus Analysis... 118
The Power of Advanced Bus Analysis.. 118
Conclusion.. 118
.... 119

Chapter 37: Advanced Bus Analysis with Waveform Automation............................ 120
Using Waveform Automation to Save Time and Increase Efficiency in Repetitive Tasks... 120
The Benefits of Waveform Automation.. 120
Utilizing Waveform Automation for Advanced Bus Analysis..................... 121
Customizing Waveform Automation to Suit Your Needs........................... 121
Improving Collaboration and Knowledge Sharing...................................... 121
In Conclusion.. 121
.... 122

Chapter 38: Power Integrity Measurements... 122
Overview of Power Integrity Tests... 123
Power Rail Measurement.. 123
Power Consumption Analysis... 123
Performing Power Integrity Tests with an Oscilloscope............................. 124
Best Practices for Power Integrity Measurements...................................... 124
Conclusion.. 124

Chapter 1: Introduction to Oscilloscopes

Basic Principles

When it comes to electronic measurement and testing, the oscilloscope is an essential tool for engineers, technicians, and hobbyists alike. It is a sophisticated, yet versatile instrument that can visually represent electrical signals in a graphical form. By displaying the changes in voltage over time, the oscilloscope provides valuable insights into the behavior and characteristics of electronic signals.

At its core, the oscilloscope uses the principle of time-domain analysis, where the horizontal axis represents time and the vertical axis represents voltage. This allows for the observation and measurement of waveforms, which are graphical representations of electronic signals. By capturing and displaying these waveforms, the oscilloscope can measure and analyze various parameters such as amplitude, frequency, and phase.

Types of Oscilloscopes

Over the years, oscilloscopes have evolved from analog to digital, and today we have a wide variety of models available in the market. Each type of oscilloscope has its own set of features and capabilities, catering to different applications and user needs. An analog oscilloscope uses a cathode-ray tube (CRT) to display waveforms in real-time. While it may not have the advanced features of a digital oscilloscope, it still remains a reliable and cost-effective option for basic circuit analysis and measurement.

On the other hand, a digital oscilloscope uses a digitizing technique to capture and display waveforms in a digital format. This allows for more advanced features such as waveform storage, automatic measurements, and data analysis. Within digital oscilloscopes, there are four main categories: digital storage oscilloscopes (DSO), digital phosphor oscilloscopes (DPO), mixed-signal oscilloscopes (MSO), and digital sampling oscilloscopes (DSO). Each type has its own unique features and capabilities, making them suitable for different applications.

The Importance of Digital Oscilloscopes

In recent years, digital oscilloscopes have become increasingly popular due to their advanced features and capabilities. With the rapid advancement of technology, electronic circuits have also become more complex, making it challenging for analog oscilloscopes to accurately measure and analyze these circuits. Digital oscilloscopes, on the other hand, can handle complex waveforms and provide more accurate measurements. They also offer a wide variety of analysis and measurement tools and can store and export data for further analysis. Additionally, digital oscilloscopes can be remotely controlled and automated, making them a valuable asset in large-scale testing and production environments. Furthermore, digital oscilloscopes are often equipped with advanced triggering options, making it easier to capture and analyze specific events and signals of interest. This is crucial for troubleshooting and debugging electronic circuits, saving time and effort for the user.

In conclusion, the advent of digital oscilloscopes has revolutionized the field of electronic testing and measurement. With their advanced features, accuracy, and versatility, they have become an essential tool in various industries and applications. In this book, we will delve deeper into the world of digital oscilloscopes, exploring their features, applications, and best practices for optimal usage. So, let's dive in and discover the endless possibilities that digital oscilloscopes offer!

Chapter 2: Evolution of Digital Oscilloscopes

Evolution from Analog to Digital

The journey of oscilloscopes began with analog technology, which used a cathode ray tube (CRT) to display waveforms. The analog oscilloscopes were bulky, expensive, and had limited functionality. However, they laid the foundation for the development of more advanced digital oscilloscopes. In the late 1990s, digital oscilloscopes started to gain popularity due to their ability to capture and store digital waveforms. The waveform was sampled, digitized, and stored in a memory buffer, allowing for more accurate and detailed analysis. With the advancement of digital technology, oscilloscopes evolved rapidly, offering higher bandwidths, faster sampling rates, and more advanced features.

Key Milestones

The digital oscilloscope market has seen many groundbreaking innovations that have shaped the way we use oscilloscopes today. Let's take a look at some of the key milestones in the evolution of digital oscilloscopes.

1974:
Tektronix introduces the first commercial digital oscilloscope, the 4000 series, which could display four waveforms simultaneously.

1983:
Hewlett-Packard (now known as Agilent Technologies) introduces the first mixed-signal oscilloscope that could display both analog and digital waveforms.

1995:
LeCroy introduces the first oscilloscope with a color display, making waveform analysis easier and more intuitive.

1999:
Tektronix introduces the first digital storage oscilloscope (DSO) with a bandwidth of 1

GHz, setting a new standard for oscilloscopes in the high-frequency range.

2003:
Rohde & Schwarz introduces the first oscilloscope with a built-in spectrum analyzer, combining two essential tools for signal analysis into one device.

2007:
Agilent Technologies creates the first oscilloscope with a touchscreen display, revolutionizing the user interface and making waveform analysis more efficient.

2017:
Keysight Technologies introduces the first 110 GHz real-time oscilloscope, setting a new record for oscilloscope bandwidth and allowing for the analysis of ultra-high-frequency signals.

Major Contributors

Numerous companies and individuals have contributed to the evolution of digital oscilloscopes. Some of the major players in the market include Tektronix, Keysight Technologies, Rohde & Schwarz, and Fluke Corporation, among others. However, a few names stand out for their significant contributions to the development of oscilloscopes. Alan Kuehne, the creator of the first digital oscilloscope at Tektronix, is considered the father of the digital oscilloscope. His invention paved the way for others to build upon and improve digital oscilloscopes. Another significant contributor is Dave Jones, a popular YouTuber and engineer who has influenced the DIY electronics community with his insightful oscilloscope reviews and tutorials.

Closing Thoughts

The evolution of digital oscilloscopes has come a long way since its humble beginnings as an analog technology. With constant developments and advancements, oscilloscopes today have become essential tools for engineers, scientists, and enthusiasts alike.

In the next chapters, we will dive deeper into the technology behind digital oscilloscopes and understand how to make the most out of their features for accurate waveform analysis. So stay tuned and get ready to explore the world of digital oscilloscopes.

Chapter 3: Understanding Oscilloscope Specifications

Bandwidth

When it comes to selecting the right oscilloscope for your needs, one of the most important specifications to consider is bandwidth. Bandwidth refers to the range of frequencies that an oscilloscope can accurately measure. It is usually measured in Hertz (Hz) or megahertz (MHz) and is typically displayed as a number followed by the unit, such as 100 MHz.

A higher bandwidth means that the oscilloscope can accurately capture and display a wider range of frequencies, making it ideal for measuring high-frequency signals. However, it is important to note that the bandwidth requirement depends on the type of signals you will be working with. For example, if you are working with digital signals with fast rise times, a higher bandwidth is essential to accurately represent the signal.

Sampling Rate

The sampling rate is another crucial specification to consider when selecting an oscilloscope. It refers to the number of samples the oscilloscope takes per second, which is then used to reconstruct the waveform on the display. The sampling rate is important because it determines how accurately the signal can be represented.

The Nyquist sampling theorem states that the sampling rate should be at least twice the frequency of the signal being measured. Therefore, if you are working with a signal that has a frequency of 100 MHz, your oscilloscope should have a sampling rate of at least 200 MHz to accurately display the signal.

Memory Depth

Memory depth is another important specification to take into consideration when selecting an oscilloscope. It refers to the amount of data points that can be stored and displayed on the oscilloscope's screen. A larger memory depth allows for a longer time period to be displayed, which is especially useful when capturing infrequent or sporadic events.

When choosing an oscilloscope, it is important to consider the type of signals you will be working with and the amount of time you need to capture. This will help determine the necessary memory depth for your specific needs.

Other Important Features

Aside from bandwidth, sampling rate, and memory depth, there are other important features to consider when selecting an oscilloscope. These include trigger types, display resolution, and signal processing capabilities. Trigger types allow the oscilloscope to capture and display signals in specific situations, such as when a specific voltage level is reached or when a specific pattern is detected. Display resolution refers to the number of pixels on the screen, which affects the accuracy of the displayed waveform. Lastly, signal processing capabilities such as waveform math functions and frequency analysis can greatly enhance the oscilloscope's capabilities for advanced measurements and analysis.

Tips for Selecting the Right Oscilloscope

With so many different oscilloscope specifications and features to consider, here are a few tips to help you select the right one for your needs:

- Determine the type of signals you will be working with and their frequency range.

- Consider the sampling rate necessary for accurate representation of the signals.

- Determine the needed memory depth for your specific needs.

- Do your research on trigger types and additional features that may be useful for your applications.

- Consider a modular oscilloscope if you need to expand its capabilities in the future.

Wrapping Up

Oscilloscope specifications play a crucial role in selecting the right tool for your measurement needs. By understanding the different aspects such as bandwidth, sampling rate, and memory depth, you can confidently choose the best oscilloscope to suit your specific applications. Remember to consider the type of signals you will be working with, as well as any additional features that may enhance your measurements and analysis. With the right tool in hand, you can confidently tackle any digital oscilloscope application with ease and precision.

Chapter 4: Choosing the Right Oscilloscope

Factors to Consider

Choosing the right oscilloscope can seem like a daunting task, especially for beginners. With so many options on the market, it's important to consider several factors before making a purchase. Here are some key factors to consider when choosing your first digital oscilloscope.

Bandwidth

The bandwidth of an oscilloscope is the range of frequencies that the scope can accurately capture. It is an important factor to consider, as it determines the highest frequency signal that the scope can accurately display. In general, it is recommended to choose an oscilloscope with a bandwidth that is at least five times higher than the highest frequency signal you plan on measuring.

Sample Rate

Sample rate is the number of times per second that an oscilloscope collects data points. This is also an important consideration, as a higher sample rate allows for more accurate representation of signals with high frequency components. It is recommended to choose an oscilloscope with a sample rate of at least two to three times higher than the highest frequency signal.

Channels

Another factor to consider is the number of channels on the oscilloscope. Most scopes will come with either two or four channels, with some higher-end models offering up to eight. The number of channels you need will depend on the types of signals you plan on measuring. If you primarily work with single-ended signals, then two channels may be sufficient. However, if you work with differential signals or need to compare multiple signals simultaneously, then a scope with more channels would be beneficial.

Memory Depth

Memory depth refers to the amount of data points that an oscilloscope can store. This is important when capturing and analyzing complex signals. The higher the memory depth, the more accurately the oscilloscope can capture and display long or complex signals. In general, it is recommended to choose an oscilloscope with a memory depth of at least 10 times the duration of the signal being measured.

Price Range

When it comes to oscilloscopes, the price range is quite broad. You can find basic oscilloscopes for under $100, while high-end models can cost thousands of dollars. As a beginner, it may be tempting to opt for a cheaper model, but keep in mind that a lower price often means sacrificing some features and specifications. It's important to strike a balance between your budget and the features you need.

Budget Friendly

If you are on a tight budget, there are still some good options for beginner oscilloscopes. Brands such as Siglent, Rigol, and Tektronix offer entry-level scopes with basic features at affordable prices. These scopes typically have lower bandwidths and sample rates, but are still suitable for basic measurements.

Mid-Range

For those willing to invest a bit more, there are mid-range options from brands such as Keysight and GW Instek. These scopes offer higher specifications and additional features like serial bus decoding and advanced triggering options. They are suitable for more advanced measurements and troubleshooting.

High-End

If you require top-of-the-line performance and features, then high-end oscilloscopes from brands like LeCroy, Teledyne LeCroy, and Rohde & Schwarz are worth considering. These scopes offer the highest bandwidths, sample rates, and sophisticated measurement capabilities. However, they come at a premium price.

Recommended Brands

When it comes to choosing the right oscilloscope, brand reputation is also an important factor to consider. Here are some recommended brands, known for their quality and reliability in the industry.

Tektronix

Founded in 1946, Tektronix is a well-respected brand in the oscilloscope market. They offer a range of scopes from basic to high-end, with a focus on innovation and cutting-edge technology. Their scopes are known for their user-friendly interfaces and accurate measurements.

Keysight Technologies

Another well-established brand, Keysight offers a wide selection of high-quality oscilloscopes. Their scopes are designed with precision and versatility in mind, making them a top choice for researchers and professionals.

Rigol

Rigol is a relatively new player in the oscilloscope market but has gained a reputation for offering affordable yet reliable scopes. Their entry-level DS1000Z series has become a popular choice for beginners and hobbyists.

Teledyne LeCroy

Known for their high-end oscilloscopes, Teledyne LeCroy has been in the business since

1964. Their scopes are known for their advanced features and capabilities, making them a top choice for professionals in fields such as telecommunications and aerospace.

Siglent

Founded in 2002, Siglent has quickly become a rising star in the oscilloscope market. They offer a range of scopes with advanced features at an affordable price point, making them a top choice for budget-conscious buyers.

Final Thoughts

Choosing the right oscilloscope can be overwhelming, but by considering the key factors mentioned above and sticking to a reputable brand, you can ensure that you have a reliable and capable instrument for your needs. As you grow in your skills and require more features, you can always upgrade to a higher-end model. With some research and consideration, finding the perfect oscilloscope for your needs is achievable.

Chapter 5: Initial Setup and Calibration

Power Supply

When setting up your digital oscilloscope, the first and most important step is to make sure it is properly connected to a stable power supply. This will ensure accurate and reliable measurements throughout your testing process. Most oscilloscopes come with a power cord that can be plugged into a standard outlet. If you are using a battery-operated oscilloscope, make sure the battery is fully charged before beginning your measurements. One thing to keep in mind is that the power supply can greatly affect the noise and interference levels in your measurements. It is recommended to use a separate, dedicated power supply for your oscilloscope to minimize any potential disturbances.

Grounding

Grounding is another crucial aspect of setting up your oscilloscope. It is important to have a good ground connection to maintain accurate measurements and avoid potential damage to your equipment. Make sure to use a designated grounding point on your oscilloscope and connect it to a known operating ground. Avoid connecting the grounding probe to any large metal objects that may introduce noise into your measurements.

Probe Connections

The probes used with an oscilloscope play a critical role in accurately capturing and measuring waveforms. Before connecting your probes, make sure they are functioning properly and have been properly compensated. When connecting your probes, ensure that they are connected to the designated input channels on your oscilloscope. Also, take care to use the appropriate probe for the signal you are measuring, whether it is an analog or digital signal.

Other Initial Setup Steps

In addition to power supply, grounding, and probe connections, there are a few other initial setup steps you should take to ensure accurate and reliable measurements from your oscilloscope. First, check the vertical and horizontal scale settings on your oscilloscope to ensure they are appropriate for the signal you are measuring. Next, adjust the trigger level to the desired voltage level. If you are measuring a repetitive signal, you can also set up a trigger delay to capture the waveform at a specific point in time. This can be useful when analyzing specific portions of a waveform.

Lastly, it is important to perform a calibration of your oscilloscope before beginning your measurements. This will ensure that your instrument is functioning properly and will provide accurate results.

In Conclusion

A properly set up and calibrated oscilloscope is essential for accurate and reliable measurements. By following these initial setup steps and paying attention to details such as power supply, grounding, and probe connections, you can ensure that your oscilloscope is ready to assist you in your testing process. Always remember to perform a calibration before beginning your measurements and refer to the manufacturer's guidelines for any specific recommendations for your particular oscilloscope model. Happy measuring!

Chapter 6: Display Controls and Menus - Your Guide to Navigating the World of Oscilloscopes

Display controls, control knobs, and menu options

As you start using your digital oscilloscope, you will find yourself faced with a multitude of display controls, control knobs, and menu options. These are the tools that will help you navigate through the vast ocean of signals and waveforms. Understanding how to use these controls effectively is essential to mastering the art of oscilloscope operation.

The Display Control Panel

The display control panel is your window into the world of signals and waveforms. This is where you'll find the controls for adjusting the display settings, such as the vertical and horizontal scale, trigger levels, and waveform persistence. The first thing you'll notice on the panel is the screen, which displays the signal and waveform you're measuring. The screen may also feature grids and markers to help you visualize the signals more clearly. One key control you'll use on the display panel is the intensity adjustment. This control allows you to adjust the brightness of the display, making the signals easier to see. You can also find controls for adjusting the position of the signal on the screen, as well as its focus and size.

Control Knobs and Buttons

Next to the display panel, you will find a series of control knobs and buttons. These are used to adjust the various settings and parameters of your oscilloscope. The main knob you'll use is the vertical position knob, which allows you to move the signal up or down on the screen to get a better view. The vertical scale knob allows you to adjust the size of the signal, while the horizontal scale knob controls the time scale of the waveform. Other important controls include the trigger source and mode buttons. These determine which signal the oscilloscope will use as a trigger and the type of trigger it will use. The trigger is an essential feature that helps you capture and stabilize the waveform for easier analysis. You can also find buttons for controlling the probes, such as the probe attenuation and probe coupling.

The Menu Options

In addition to the physical controls on the oscilloscope, you can also access various menu options through the display panel. These menus allow you to adjust more advanced settings and access additional features and functions. Some common menu options include the trigger menu, which allows you to fine-tune the trigger settings, and the measurement menu, where you can perform advanced measurements such as frequency, amplitude, and rise time.

One important feature that you can access through the menus is the waveform persistence or persistence control. This feature allows you to visualize multiple signals or waveforms simultaneously, making it easier to compare and analyze them.

Understanding the Hierarchical Menus

Most modern digital oscilloscopes have hierarchical menus, meaning that they have multiple layers of sub-menus that you can navigate through to access different features and options. It may seem overwhelming at first, but once you understand the structure of the menus, you'll find that they are designed with efficiency and ease of use in mind. The top-level menu contains the most frequently used options, such as the vertical and horizontal scales, trigger settings, and time base settings. Within each of these options, you'll find sub-menus that allow you to adjust the parameters in more detail. It's important to take your time and familiarize yourself with these menus to get the most out of your oscilloscope.

Navigating the Menus with Soft Keys and Knobs

Most digital oscilloscopes come with soft keys and knobs that allow you to navigate through the menus more efficiently. The soft keys are buttons on the display panel that correspond to various menu options, and the knobs can be pushed or rotated to adjust the settings for the selected option.

One helpful tip for navigating the menus is to use the auto button. This button automatically selects the most commonly used menu options, making it a useful tool for beginners. Additionally, many oscilloscopes come with a help button that provides information and tips on how to use the menus effectively.

Take Your Time and Experiment

As with any new technology, it's essential to take your time and experiment with the various display controls, knobs, and menu options. Don't be afraid to play around with the settings and get a feel for how they affect the signal and waveform. The more you practice, the more comfortable you'll become with using these controls, and the more you'll learn about the signals and circuits you're measuring.

In Conclusion

The display controls, control knobs, and menu options on your oscilloscope are the keys to unlocking the potential of this incredible piece of equipment. By familiarizing yourself with these controls and taking the time to explore their capabilities, you'll be well on your way to becoming a skilled oscilloscope operator. So don't be afraid to dive in and discover all that your oscilloscope has to offer.

Chapter 7: Voltage, Frequency, and Time Measurements: An Introduction to Basic Waveform Analysis Techniques

Voltage Measurements: Understanding the World of Oscilloscopes

At the heart of every oscilloscope is the measurement of voltage. It is a fundamental property of electronic signals and understanding how to properly measure and analyze voltage is crucial to getting the most out of your oscilloscope. But what exactly is voltage? Voltage is defined as the electrical potential difference between two points, and it is typically measured in volts (V). It is often referred to as the "pressure" or "force" that drives the flow of electrons in a circuit. In other words, voltage is what causes current to flow. Without it, there would be no movement of electrons and therefore no electronic devices would function.

In order to measure voltage with an oscilloscope, a probe is typically attached to the signal being measured. The probe converts the voltage into a small electrical signal that is then amplified and displayed on the oscilloscope screen. Different probes have different attenuation factors, which affect the accuracy of the voltage measurement. It is important to use the correct probe with the correct attenuation for accurate voltage measurements.

Frequency Measurements: The Music of the Oscilloscope World

Frequency is another key measurement when it comes to oscilloscopes. Put simply, frequency is the rate at which a particular event or signal occurs. It is often measured in hertz (Hz) and is directly related to time and the number of cycles or repetitions of a signal in a given period. For example, in music, frequency is associated with the pitch of a note. The higher the frequency, the higher the pitch. In electronics, frequency is typically used to measure the speed or rate of a signal, such as the speed of a processor or the frequency of an AC voltage.

Oscilloscopes have the ability to measure frequency by counting the number of cycles or peaks of a signal within a given time period. The faster the number of cycles, the higher the frequency. With the use of advanced triggering and math functions, you can also measure the frequency of specific portions of a signal, allowing you to analyze different components or disturbances within a single signal.

Time Measurements: Capturing the Essence of Signals

Another essential measurement for oscilloscopes is time. Time can be defined as the duration of an event or signal, and it is typically measured in seconds (s), milliseconds (ms), or microseconds (μs). To accurately measure time, oscilloscopes use an internal clock or timebase, which is essentially a master oscillator that controls the horizontal sweep of the display. This sweep corresponds to time, and with the use of time cursors or measurements, you can precisely measure the duration of a signal or event.

With the ability to make both voltage and time measurements, oscilloscopes allow you to fully capture the essence of electronic signals and analyze them in detail.

Basic Waveform Analysis Techniques: The Art of Reading and Interpreting Waveforms

One of the most critical skills to master when using an oscilloscope is the ability to analyze and interpret waveforms. A waveform is a graphical representation of a signal, with time being represented on the horizontal axis and voltage on the vertical axis. By closely examining the shape and characteristics of a waveform, you can gain valuable insights into the behavior of a signal and pinpoint any issues or abnormalities. There are several key elements to keep in mind when analyzing waveforms. The first is the amplitude, which is the vertical height of the waveform and represents the strength of the voltage. The second is the frequency, which is the number of cycles or repetitions of the waveform within a given time period. The third is the phase, which is the timing relationship between two waveforms.

Additionally, understanding the different types of waveforms (sine, square, triangular, etc.) and being able to identify key features such as rise/fall times, period, and duty cycle can greatly aid in waveform analysis.

Conclusion

Voltage, frequency, and time measurements, along with basic waveform analysis techniques, are the foundational aspects of using an oscilloscope. By mastering these skills, you can accurately measure and analyze electronic signals and troubleshoot any issues that may arise. As you continue to delve deeper into the world of oscilloscopes, you will discover more advanced techniques and applications for these measurements, but it is essential to have a solid understanding of the basics first. With patience and practice, you can become an expert in voltage, frequency, and time measurements, and unlock the full potential of your oscilloscope.

Chapter 8: Saving and Recalling Waveforms

Oscilloscopes are powerful tools for capturing and analyzing electronic signals. However, all the effort put into capturing the waveform of interest can be lost if the data is not saved or is accidentally overwritten. This is where the importance of saving and recalling waveforms comes into play. In this chapter, we will explore the various methods of saving and recalling waveforms on a digital oscilloscope and provide tips for organizing saved waveforms.

How to Save and Recall Waveforms

There are multiple ways to save and recall waveforms on a digital oscilloscope. The most common method is to save the waveform to the internal memory of the oscilloscope. This memory can vary in size depending on the model and manufacturer, but typically it can store hundreds or even thousands of waveforms. When saving a waveform, it is important to give it a meaningful name and a description if possible. Another way to save waveforms is by using a USB drive. Most modern oscilloscopes come equipped with a USB port, making it easy to transfer data to and from the oscilloscope. By saving waveforms to a USB drive, it is possible to store a large number of waveforms without worrying about limited memory on the oscilloscope. This is especially useful when working on multiple projects or when sharing data with others.

Some oscilloscopes also have the ability to save waveforms to a computer via Ethernet or Wi-Fi connectivity. This is particularly helpful when working in a laboratory setting where the oscilloscope may be located in a different room or on a different floor. With this feature, waveforms can be saved directly to a computer for further analysis or sharing with colleagues.

Tips for Organizing Saved Waveforms

Organizing saved waveforms is crucial for efficiency and ease of access. With the potential to save hundreds of waveforms, it can quickly become overwhelming and difficult to find the desired waveform if they are not organized properly. Here are some tips for organizing saved waveforms:

- Create folders: Most oscilloscopes allow for the creation of folders within the internal memory or a USB drive. These folders can be named according to project, date, or any other distinguishing feature. By categorizing waveforms into different folders, it becomes much easier to locate them later on.

- Use descriptive names: When saving a waveform, it is important to give it a meaningful name that accurately describes the signal and any relevant settings or conditions. This makes it easier to identify the waveform later on, especially when working on multiple projects at once.

- Use descriptions: Many oscilloscopes also allow for the input of a waveform description. This can include relevant information such as the equipment used, the purpose of the measurement, and any additional notes. Having this information readily available can be useful when reviewing saved waveforms at a later time.

- Utilize the tagging feature: Some oscilloscopes have a tagging feature that allows for the addition of tags to saved waveforms. Tags can be used to group waveforms with similar characteristics or to denote certain parameters or conditions. This can be helpful for quick and easy filtering of waveforms.

- Delete unused or duplicate waveforms: As it is with any type of data, it is important to regularly clean up and delete any unused or duplicate waveforms. This not only helps with organization, but it also frees up memory on the oscilloscope for future use.

Proper organization of saved waveforms not only saves time but also ensures that important data is not accidentally deleted or lost. By following these tips, it becomes much easier to navigate and manage large numbers of saved waveforms.

Conclusion

Saving and recalling waveforms is an essential skill for effectively using an oscilloscope. By utilizing the various methods of saving and organizing waveforms, it becomes much easier to review and analyze captured data. Whether it is through the internal memory, USB drive, or computer connectivity, always be mindful of properly labeling and categorizing saved waveforms to ensure efficiency and accurate data retrieval. With this knowledge, you are well on your way to becoming a master of digital oscilloscopes.

Chapter 9: Understanding and Mastering Triggering and Capture Modes

Oscilloscopes are fantastic tools for capturing and analyzing waveforms, but they are only as good as their triggering capabilities. This is where the heart of the oscilloscope lies, as it allows you to capture and isolate specific waveforms for analysis. In this chapter, we will dive into the different trigger types and how to use them effectively for specific applications.

Different Trigger Types and How to Use Them for Specific Applications

There are various trigger types available on oscilloscopes, and choosing the right one for your application is crucial for accurate waveform capture. The most commonly used trigger types include edge, pulse, video, and pattern. Let's look at each of these in more detail and understand how they can be used to make the most of your oscilloscope.

Edge Triggering

Edge triggering is the most basic and commonly used trigger type. It allows you to trigger the oscilloscope when an input signal crosses a specific voltage threshold. The edge type can be set to either rising or falling edge, depending on the desired trigger point. This type of trigger is useful for capturing transient events and signals with fast rise times. To use edge triggering effectively, it is crucial to set the trigger level accurately. This level should be set just below or above the expected voltage level of the signal. This ensures that the trigger occurs at the desired point and gives a stable waveform display.

Pulse Triggering

Pulse triggering is similar to edge triggering but has an added trigger holdoff feature.

This allows the oscilloscope to ignore additional triggers within a specified period after the trigger event. It is useful for eliminating repetitive waveforms and capturing isolated pulses within a train of pulses. To use pulse triggering, it is essential to set the holdoff time correctly, depending on the pulse width you want to capture. This ensures that the oscilloscope does not trigger on unwanted pulses and gives a clean display of the desired pulse.

Video Triggering

Video triggering is a specialized type of trigger used for capturing video signals. It is designed to allow for the synchronization of the vertical and horizontal deflection of the oscilloscope display with the video signal. This ensures a stable and consistent display of the video signal.

To use video triggering, the horizontal scale should be set to match the frequency of the video signal. This allows for a proper display of the signal, and the trigger level should be set to the center of the waveform for stable triggering.

Pattern Triggering

Pattern triggering is useful for capturing signals with specific patterns or sequences. It allows you to trigger the oscilloscope when a particular pattern is detected within the signal. This type of trigger is commonly used for serial data communication signals and digital circuit debugging. To use pattern triggering, you need to define the expected pattern in the oscilloscope's trigger menu accurately. This ensures that the oscilloscope only triggers when the desired sequence is detected within the signal.

Advanced Triggering Techniques

In addition to the basic triggering techniques mentioned above, many advanced triggering capabilities are available in modern oscilloscopes. These include pulse width, runt, window, and logic triggering. These advanced techniques provide more precise and specialized triggering options for specific applications, such as digital communication and power analysis.

It is essential to explore and understand these advanced triggering techniques to get the most out of your oscilloscope and accurately capture and analyze waveforms.

Tips for Effective Triggering

Having the right trigger setup is crucial for accurate waveform capture, but there are a few key tips to keep in mind to ensure the best results. These include:

- Setting the trigger level accurately: As mentioned earlier, the trigger level should be set just below or above the expected voltage level of the signal. This ensures stable triggering and a clear display of the waveform.

- Adjusting the trigger holdoff: For pulse and video triggering, it is essential to set the trigger holdoff time accurately to avoid unwanted triggering and get a clean display of the desired signal.

- Experimenting with different trigger types: Don't be afraid to try out different triggering techniques to see which one works best for your specific application. Some signals may require a different trigger type than others, so it is essential to have a good understanding of all the options available.

Mastering Triggering and Capture Modes for Accurate Measurements

In conclusion, mastering triggering and capture modes is crucial for getting accurate measurements and analysis of waveforms with an oscilloscope. With the right trigger setup and a good understanding of the different techniques available, you can make the most of your oscilloscope and troubleshoot circuits with ease. So don't be afraid to experiment and learn about the various triggering options to become a pro at capturing and analyzing waveforms.

Chapter 10: Advanced Measurements and Their Role in Troubleshooting

Digital oscilloscopes are essential tools for troubleshooting and analyzing electronic circuits. They provide a wealth of information about signals, such as their amplitude, frequency, and timing. However, sometimes basic measurements are not enough to fully understand a complex signal or isolate an issue. This is where advanced measurements come into play. In this chapter, we will explore the RMS, peak-to-peak, and other advanced measurements and how they can be used for troubleshooting purposes.

RMS (Root Mean Square) Measurements

One commonly used measurement is Root Mean Square, or RMS for short. RMS is a mathematical calculation used to determine the effective voltage or current of a signal. This is important, as some signals may appear to have a lower amplitude, but may actually be delivering more power than a higher amplitude signal. RMS measurements are particularly useful for AC signals, where the voltage or current is constantly changing direction.

In order to use the RMS measurement on your oscilloscope, the signal must be a single repetitive waveform. You can set the oscilloscope to automatically calculate the RMS value, or you can use the "measure" function to manually calculate it for a specific part of the waveform. This can be helpful in troubleshooting issues with power delivery, as it gives you a better understanding of the actual power being delivered by the signal.

Peak-to-Peak Measurements

Peak-to-peak measurements are another useful tool for troubleshooting. This measurement gives you the difference between the highest and lowest points of a waveform. This can be helpful in identifying issues such as noise and interference, as well as understanding the range of a signal. By using the peak-to-peak measurement, you can determine the maximum and minimum values of a signal and compare them to the expected values.

You can also use this measurement to analyze the amplitude of a signal and its deviation from its expected value. In addition, peak-to-peak measurements can be useful for identifying signal distortions and irregularities that may be causing issues in your circuit.

Other Advanced Measurements

Digital oscilloscopes offer a wide range of advanced measurements that can help in troubleshooting complex issues. Some of these measurements include rise time, fall time, pulse width, and duty cycle. Rise time measures the time it takes for a signal to transition from its low value to its high value, while fall time measures the transition from high value to low value. These measurements can be useful in identifying issues with fast-changing signals.

Pulse width measures the time between the start and end of a pulse in a signal, while duty cycle measures the ratio of the pulse width to the total period of the signal. These measurements are useful for analyzing digital signals and identifying issues such as clock skew and signal degeneration.

Using Advanced Measurements for Troubleshooting

With the plethora of advanced measurements available on digital oscilloscopes, it can be overwhelming to know which ones to use for troubleshooting purposes. Firstly, it is important to understand the different measurements and what they can tell you about your signal. It is also worth noting that these measurements should not be used in isolation, but rather in combination with other measurements and your own analysis. When facing a troubleshooting issue, start by evaluating the signal using the basic measurements such as amplitude and frequency. If those measurements do not provide enough information, then start experimenting with advanced measurements. Remember to always refer back to the specifications of your circuit and compare the results of your measurements to expected values.

Conclusion

In this chapter, we have explored the various advanced measurements available on digital oscilloscopes and their role in troubleshooting complex electronic circuits. From RMS to peak-to-peak and other measurements, these tools can provide valuable insights into the behavior of signals and help identify issues that may not be apparent with basic measurements. It is important to use these measurements in conjunction with other troubleshooting techniques and to also have a thorough understanding of the expected values. With proper utilization, advanced measurements can be invaluable in solving even the most complex issues.

Chapter 11: Probes and Probe Compensation

Types of Probes

When using an oscilloscope, the probe you choose can greatly affect the accuracy and quality of your measurements. There are a variety of probes available, each designed for specific types of signals and applications. Some common types of probes are:

Passive Probes

These are the most basic type of probes and are typically included with oscilloscopes. They consist of a coaxial cable with a metal tip at one end and a BNC connector at the other. Passive probes are suitable for low-frequency signals and can handle voltages up to 500V.

Active Probes

Active probes are more advanced and have their own power supply and amplification circuitry. They provide better signal fidelity and are suitable for high-frequency signals. However, they are also more expensive than passive probes.

Current Probes

Current probes are used for measuring current signals and require the use of a 50-ohm termination adapter. They can provide higher accuracy and bandwidth than other types of probes, making them ideal for power measurements.

Differential Probes

Differential probes are designed for measuring differential signals, such as those found in high-speed serial data communications. They can provide better common-mode rejection and are suitable for measuring fast rise-time signals.

Probe Specifications

When choosing a probe, there are several important specifications to consider. These include:

Bandwidth

The bandwidth of a probe determines the range of frequencies it can accurately measure. It is typically measured in MHz or GHz and should be chosen based on the frequency range of the signals you will be measuring.

Rise Time

Similar to bandwidth, rise time is a measure of how quickly a probe can respond to a sudden change in a signal. It is typically measured in picoseconds (ps) and is directly related to the probe's bandwidth.

Input Impedance

Probe input impedance, measured in ohms, is the amount of load the probe places on the circuit being measured. It is crucial to choose a probe with an input impedance that matches the input impedance of your oscilloscope for accurate measurements.

Attenuation Ratio

The attenuation ratio of a probe determines how much the measured signal is scaled down for display on the oscilloscope screen. Most probes have an attenuation ratio of 1:1 or 10:1, with 10:1 being the most common. This means that a 10:1 probe will measure a 10 times larger voltage than it displays on the screen.

Tips for Probe Compensation

Probing a circuit can introduce loading effects and influence the measured results. To compensate for these effects, it is important to properly calibrate and adjust the probe before taking measurements. Here are some tips for successful probe compensation:

Check the Ground Lead

Ensure that the probe's ground lead is properly connected and that there are no broken or loose connections. A poor ground connection can affect the accuracy of your measurements.

Use the Correct Attenuation Setting

Using the correct attenuation setting for your probe is crucial for accurate measurements. If you are unsure, it is always better to err on the side of using a lower attenuation ratio.

Compensate for Probe's Capacitive Loading

Because the probe's capacitance can affect the measured signal, most probes come with a small compensation tool to adjust for this effect. Follow the manufacturer's instructions to properly compensate for your specific probe.

Minimize Noise and Interference

Proper grounding and shielding can help minimize noise and interference, which can affect the accuracy of your measurements. Using a probe with a lower input impedance can also help reduce noise.

Refer to the Oscilloscope Manual

If you are still unsure about how to properly compensate your probe, refer to the oscilloscope's manual. It will provide specific instructions and guidance for your particular model.

In Conclusion

Choosing the right probe and properly compensating it can greatly improve the accuracy and quality of your oscilloscope measurements. Consider the type of signals and applications you will be working with and carefully select a probe with the appropriate specifications. Remember to always follow the manufacturer's instructions and refer to the oscilloscope manual for any questions. With the right probe and careful compensation, you can confidently make accurate measurements and troubleshoot electronic circuits with ease.

Chapter 12: Troubleshooting with Oscilloscopes

Common Troubleshooting Techniques using Oscilloscopes

As a beginner to the world of oscilloscopes, it can be overwhelming to figure out how to troubleshoot issues with your circuit. Fortunately, oscilloscopes have many features that can aid in identifying problems and finding solutions. Here are some common techniques that can help you troubleshoot using your oscilloscope.

Finding Ground Loops

Ground loops can cause unwanted noise and interference in your circuit, and can be a headache to troubleshoot. However, with the help of your oscilloscope, you can easily locate and eliminate ground loops. One approach is to use the 'ground reference' or 'floating ground' feature on your oscilloscope. This allows you to select a separate ground reference point, which can help you identify where any ground loops may be occurring. You can then use this information to isolate the issue and make necessary adjustments to your circuit. Another technique is to use a 'differential probe' or 'current probe' with your oscilloscope. These probes allow you to measure the voltage drop across specific points in your circuit, which can help pinpoint where a ground loop is occurring. With this information, you can identify potential sources of noise and take steps to eliminate it.

Signal Distortions

Signal distortions can occur due to a variety of factors, such as noise, interference, or improper connections. Troubleshooting signal distortions requires a keen eye and attention to detail. Luckily, oscilloscopes have features that can help make this process easier. One commonly used technique is to use the 'advanced trigger' feature on your oscilloscope. This allows you to set triggering conditions based on specific signal characteristics, such as amplitude, frequency, or pulse width. With this feature, you can capture and analyze the distorted waveform, and gain insight into the root cause of the issue.

Another useful tool is the 'math function' feature on your oscilloscope. This allows you to perform mathematical operations on waveforms, such as addition, subtraction, or division. By comparing the distorted signal to a known good signal, you can use these math functions to identify any discrepancies and troubleshoot accordingly.

Takeaways

Troubleshooting with oscilloscopes may seem like a daunting task, especially for beginners. However, with the right techniques and features, you can easily identify and resolve issues in your circuit. Remember to use the 'ground reference' or 'floating ground' feature to locate ground loops, and the 'advanced trigger' and 'math function' features to troubleshoot signal distortions. By taking advantage of these and other tools within your oscilloscope, you can become a skilled troubleshooter and ensure the success of your circuit designs.

Chapter 13: Advanced Features and Applications

When it comes to digital oscilloscopes, there is much more than meets the eye. While they are incredibly useful tools for basic measurements and troubleshooting, they also have advanced features that can take your work to the next level. In this chapter, we will explore some of these features and applications in detail, including FFT analysis, serial protocol decoding, and other advanced functions.

FFT Analysis

FFT analysis, or Fast Fourier Transform analysis, is a powerful tool that allows you to break down a complex signal into its individual frequency components. This is especially useful for analyzing signals in the frequency domain, such as audio and radio signals. With FFT analysis, you can see how much of each frequency component is present in the signal and identify any harmonic distortions or noise.

To use FFT analysis on your digital oscilloscope, you will need to select the FFT mode on your display. Then, you can adjust the frequency range and scale to focus on specific frequencies of interest. The resulting FFT plot will show the amplitude of each frequency component, allowing you to identify any issues or troubleshoot your circuit.

Serial Protocol Decoding

One of the most common uses for digital oscilloscopes is for analyzing digital signals. With the rise of digital communication and serial protocols, such as SPI, I2C, and UART, it has become essential to have the capability to decode and analyze these signals. Fortunately, many modern oscilloscopes have built-in decoding features that can automatically interpret and display serial data as waveforms. To use serial protocol decoding, you will need to connect your oscilloscope probe to the data signal and select the appropriate protocol on your oscilloscope. The resulting waveform will show the decoded data in a human-readable format, with markers indicating the start and end of each data packet. This can be incredibly useful for troubleshooting communication issues or analyzing the performance of your digital circuits.

Other Advanced Features and Applications

In addition to FFT analysis and serial protocol decoding, digital oscilloscopes have many other advanced features and applications that can benefit your work. Some examples include:

- Mask testing: With mask testing, you can define a specific waveform pattern as a reference and compare new waveforms against it to identify any deviations. This is useful for ensuring signal integrity and identifying anomalies.
- Math functions: Many oscilloscopes have advanced math capabilities that allow you to perform mathematical operations on waveforms. This can be useful for calculating frequency, phase, and more.
- Advanced triggering: In addition to the basic trigger modes, some oscilloscopes have advanced triggering options, such as logic triggering and runt triggering. These can help you capture and analyze specific events in your waveform.
- Automated measurements: Some oscilloscopes have features that allow you to automate measurements and save time. For example, you can set up your scope to automatically measure voltage, frequency, and other parameters and display them in a table.

Overall, these advanced features and applications can help you gain a deeper understanding of your circuit's performance and troubleshoot issues more efficiently. It is worth exploring these functions on your digital oscilloscope and understanding how they can aid your work.

Conclusion

In this chapter, we have delved into some of the advanced features and applications of digital oscilloscopes. We covered FFT analysis and its usefulness in understanding frequency components, serial protocol decoding for analyzing digital signals, and other advanced functions such as mask testing, math operations, advanced triggering, and automated measurements. These capabilities provide a more comprehensive understanding of your circuit's performance and can help you troubleshoot and identify issues quickly. With these tools at your disposal, you can take your work to the next level and become more efficient and effective in your tasks.

Chapter 14: Saving and Exporting Data from Oscilloscopes

The Importance of Saving and Exporting Data

In the world of digital oscilloscopes, the ability to save and export data is crucial. As engineers and scientists, we rely on data to validate our theories, troubleshoot problems, and make informed decisions. An oscilloscope is a powerful tool that captures and displays electrical signals in the form of waveforms, making it an essential device in any lab or workshop. However, the ability to save and export this data amplifies the usefulness of an oscilloscope, allowing us to analyze and manipulate the captured signals in different ways.

Options for Saving and Exporting Data

Digital oscilloscopes provide a variety of options for saving and exporting data. These options include CSV (Comma Separated Values) files, waveform image files, and other file formats. Each of these formats serves a specific purpose, and understanding how to use them can greatly enhance the productivity of your work.

CSV Files

CSV files are a popular option for saving and exporting data from oscilloscopes. These files contain a series of data points in a plain text format, separated by a delimiter such as a comma or a semicolon. The advantage of using CSV files is that they can be easily opened and manipulated in spreadsheet software such as Microsoft Excel or Google Sheets. This allows for further analysis and manipulation of the captured data, giving you more control and flexibility in your work.

Waveform Image Files

Apart from CSV files, oscilloscopes also have the capability to save waveforms as image files. These image files can be in various formats such as BMP, PNG, or JPEG. The advantage of using waveform image files is that they capture the visual representation of the waveform, making it easy to share and document your results. These images can be inserted into reports, presentations, or any other documents, making it a convenient way to showcase your findings.

Other File Formats

Depending on the make and model of your oscilloscope, you may also have the option to save and export data in other file formats such as MATLAB, LabVIEW, or TDMS. These formats are more commonly used in specialized software and can provide additional analysis and processing capabilities. It is essential to familiarize yourself with the available options and understand which format will best suit your needs.

Tips for Efficient Saving and Exporting

To save and export data efficiently from your oscilloscope, here are some tips to keep in mind:

- Before saving or exporting, be sure to properly label your data for easy identification later.
- If using waveform image files, make sure to adjust the scale and settings to capture the necessary details of the signal.
- Ensure that your data is captured and recorded correctly before saving or exporting.
- Organize your data into folders or use a naming convention to keep track of multiple files.
- If using a USB drive to save and export data, make sure it is properly formatted and compatible with your oscilloscope.

In Conclusion

The ability to save and export data from an oscilloscope may seem like a minor feature, but its importance cannot be overstated. It allows us to further analyze and manipulate the captured signals, share our findings, and document our work. As we continue to push the boundaries of technology, the importance of saving and exporting data will only increase. So, make sure to familiarize yourself with the options available and use them efficiently to make the most out of your digital oscilloscope.

Chapter 15: Best Practices for Using Remote Control Interfaces and Automation in Digital Oscilloscopes

Remote control interfaces and automation have become essential tools for efficient and accurate measurements in modern digital oscilloscopes. By using interfaces like USB and Ethernet, and automating measurements with scripts, users can save time and streamline their workflow, allowing for even more possibilities in their oscilloscope usage. In this chapter, we will explore the best practices for utilizing remote control interfaces and automation in digital oscilloscopes.

The Power of Remote Control Interfaces

Remote control interfaces, such as USB and Ethernet, have greatly expanded the capabilities of modern digital oscilloscopes. These interfaces allow for easy connection between the oscilloscope and a computer, enabling seamless communication and control. With the use of remote control interfaces, users can remotely access and control their oscilloscope from anywhere in the world, making troubleshooting and monitoring tasks more efficient and convenient.

Convenient and Efficient Automation

Automation has revolutionized the way measurements are taken with oscilloscopes. With the ability to automate processes, tasks can be completed faster and more accurately, freeing up valuable time for other important tasks. Automation eliminates the potential for human error and enables repeatable measurements for improved accuracy. This feature is especially useful when taking repetitive measurements, allowing for consistent results every time.

Using Remote Control Interfaces and Automation Together

The combination of remote control interfaces and automation takes oscilloscope usage to a whole new level. With remote control interfaces, users can easily access and control their oscilloscope from a computer, while automation allows for measurement tasks to be completed efficiently and accurately without manual intervention. This dynamic duo allows for a hands-off approach, giving users the freedom to multitask and complete other important tasks while their measurement is being carried out.

Streamlining Your Workflow

Utilizing remote control interfaces and automation not only makes measurement tasks faster and more accurate, but it also streamlines the workflow of using an oscilloscope. With the ability to remotely control the oscilloscope and automate measurements, tasks can be completed with greater efficiency and minimal interruption. This allows users to focus on various aspects of their measurement and analysis without being tied down to the oscilloscope.

Best Practices for Using Remote Control Interfaces and Automation

To make the most out of remote control interfaces and automation, there are some best practices to keep in mind. First, it is important to have a good understanding of the oscilloscope's capabilities and the commands available for remote control. This will ensure that all measurements can be performed accurately and without any issues. It is also important to have a reliable and stable connection between the oscilloscope and the computer. This will guarantee smooth communication between the two devices, allowing for efficient remote control and automation. When it comes to automating measurements, it is crucial to carefully create and test scripts before running them for actual measurements. This will help avoid any errors and ensure accurate and reliable results.

Another best practice is to regularly update the oscilloscope firmware and software to ensure compatibility and the latest features for remote control and automation.

The Future of Remote Control Interfaces and Automation in Oscilloscopes

As technology continues to advance, so will remote control interfaces and automation in oscilloscopes. With more advanced capabilities and features, users can expect even more convenience and efficiency in their oscilloscope usage. The integration of artificial intelligence and machine learning in oscilloscopes will further enhance the accuracy and capabilities of remote control and automation.

In Conclusion

Remote control interfaces and automation have become essential tools for maximizing the capabilities and efficiency of modern digital oscilloscopes. By utilizing these features, users can save time, streamline their workflow, and improve the accuracy of their measurements. With the potential for further advancements in technology, the future looks promising for remote control interfaces and automation in oscilloscopes.

Chapter 16: Tips for Maintaining and Calibrating Your Oscilloscope for Accurate Measurements

Tips for Regular Maintenance

Proper maintenance of your oscilloscope is essential for continuous and accurate performance. Here are some tips to keep your oscilloscope in good working condition:

1. Read the User Manual

The user manual for your oscilloscope contains important information on maintenance and troubleshooting. Make sure to read it thoroughly before attempting any maintenance procedures.

2. Keep It Clean

Dust and debris can build up inside your oscilloscope and affect its performance. Regularly clean the exterior and interior of your oscilloscope using a soft, lint-free cloth.

3. Check the Connections

Loose or damaged connections can lead to inaccurate measurements. Inspect and tighten all connections, including probes, cables, and power cords, to ensure a secure and reliable connection.

4. Inspect for Damage

Check for any visible signs of damage or wear on the exterior of your oscilloscope. If you notice any, it may be time to replace the damaged parts or send your oscilloscope for repairs.

5. Perform Firmware Updates

Manufacturers often release firmware updates for their oscilloscopes to improve performance and fix any bugs. Check for updates regularly and perform them as needed.

Tips for Calibration

A properly calibrated oscilloscope is crucial for accurate measurements. Here are some tips for ensuring your oscilloscope is calibrated correctly:

1. Understand Calibration Terms

Familiarize yourself with the different calibration terms used in your oscilloscope, such as offset, gain, and timebase. This will help you understand the calibration process better.

2. Use Calibration Signals

Many oscilloscopes come with built-in calibration signals that can be used to calibrate the instrument. These signals are specifically designed for accurate and precise calibration.

3. Check the Vertical and Horizontal Scales

Ensure that both the vertical and horizontal scales are set to the correct values during calibration. This will help in accurately representing waveforms on the display.

4. Perform Regular Calibration Checks

Even after calibration, it is important to perform regular calibration checks to ensure accurate measurements. Follow the steps outlined in your user manual for these checks.

Tips for Troubleshooting Calibration Issues

Despite regular maintenance and calibration, you may encounter some calibration issues with your oscilloscope. Here are some tips for troubleshooting these issues:

1. Inspect the Probes

Faulty or malfunctioning probes can cause incorrect measurements. Inspect the probes for any visible damage and replace them if necessary.

2. Check for Loose Connections

Loose connections can also cause calibration issues. Make sure all probes and cables are properly connected and tightened.

3. Perform a Factory Reset

If you are still experiencing calibration issues, try performing a factory reset on your oscilloscope. This will restore all settings to their original values and can often solve calibration problems.

4. Send for Professional Calibration

If all else fails, it may be time to send your oscilloscope to a professional for calibration. Manufacturers often offer calibration services or can recommend a trusted third-party calibration company.

Conclusion

Maintaining and calibrating your oscilloscope is essential for accurate measurements. By following these tips, you can ensure your oscilloscope is functioning properly and providing reliable results. Regular maintenance and calibration checks will help prolong the lifespan of your oscilloscope and ensure its accuracy for years to come. We hope this chapter has been informative and helpful in your journey towards understanding and utilizing digital oscilloscopes.

Chapter 17: Troubleshooting Common Oscilloscope Issues

Oscilloscopes are powerful tools for analyzing and troubleshooting electrical circuits, but like any piece of equipment, they can encounter problems from time to time. Whether you're a beginner or an experienced user, it's important to know how to troubleshoot common issues that may arise with your oscilloscope. In this chapter, we will discuss some tips for solving common problems, such as noise and display issues, that you may encounter while using your oscilloscope.

Tips for Solving Noise Issues

Noise can be a significant problem when using an oscilloscope to measure and analyze electrical signals. It can significantly affect the accuracy of your measurements and make it difficult to capture a clear waveform. Here are some tips for minimizing noise and interference while using your oscilloscope:

1. Use shielded cables: Shielded probes and cables can help reduce interference from external sources. Make sure to use high-quality cables that are properly shielded to minimize noise.

2. Avoid long ground leads: Long ground leads can act as antennas and pick up unwanted noise. Try to use the shortest ground lead possible, and keep it as close as possible to the signal being measured.

3. Use a low-pass filter: Most oscilloscopes have a built-in low-pass filter that can help reduce high-frequency noise. You can also use an external low-pass filter if your oscilloscope does not have this feature.

4. Turn off noisy equipment: If you're experiencing a lot of noise, try turning off any nearby equipment that may be causing interference. This can include cellular phones, computers, and other high-frequency devices.

Tips for Solving Display Issues

Display issues are another common problem that oscilloscope users may encounter. Inaccurate or distorted waveforms can make it challenging to analyze and troubleshoot circuits. Here are some tips for solving display issues with your oscilloscope:

1. Check the probe compensation: Improperly compensated probes can cause incorrect measurements and distorted waveforms. Make sure to properly adjust the probe compensation to match the frequency of the signal being measured.

2. Adjust the vertical and horizontal scales: Incorrectly set vertical and horizontal scales can also lead to distorted waveforms. Make sure to set the scales according to the signal being measured to get an accurate representation.

3. Use averaging: If you're seeing a lot of noise on the display, try using the averaging function on your oscilloscope. This can help smooth out the display and make it easier to analyze the waveform.

4. Calibrate your oscilloscope: Regularly calibrating your oscilloscope is essential to ensure accurate measurements. Check the calibration periodically and make any necessary adjustments to keep your oscilloscope functioning properly.

Tips for Solving Power Issues

Power issues can also cause problems with your oscilloscope's performance. Here are some tips for troubleshooting power issues:

1. Check the power supply: Make sure your oscilloscope is receiving a steady and sufficient power supply. A fluctuating or inadequate power supply can affect the accuracy of your measurements.

2. Use a power conditioner: If you're experiencing power fluctuations in your electrical circuit, using a power conditioner can help stabilize the power supply and improve measurement accuracy.

3. Check the grounding: Faulty grounding can cause issues with your oscilloscope's

performance. Make sure all grounding connections are secure and inspect them for any damage.

4. Use a stable surface: Oscilloscopes are sensitive instruments, and vibrations can affect their accuracy. Make sure to place your oscilloscope on a stable surface to avoid any disturbances or vibrations.

Tips for Solving Triggering Issues

Triggering is a critical function of any oscilloscope, and any issues with it can significantly impact the accuracy of your measurements. Here are some tips for troubleshooting triggering issues:

1. Adjust the trigger level and slope: Incorrectly set triggering parameters can result in incorrect measurements and waveforms. Make sure to set the trigger level and slope according to the signal being measured.

2. Use proper trigger coupling: Choosing the correct trigger coupling is essential to get an accurate measurement. Make sure to select the appropriate coupling for the type of signal being measured.

3. Check the trigger bandwidth: If you're experiencing issues with your trigger, make sure to check the trigger bandwidth. A trigger bandwidth that is too narrow can cause inaccurate measurements.

4. Use advanced triggering features: Advanced trigger features, such as pulse width and window triggering, can help capture specific events in a complex waveform. Familiarize yourself with these features and use them when necessary to get a more detailed analysis of your signal.

Conclusion

With the help of these tips, you can troubleshoot and solve common issues that you may encounter while using your oscilloscope. Remember to regularly check and maintain your oscilloscope to ensure its optimal performance. And most importantly, keep practicing and experimenting with your oscilloscope to become a master at troubleshooting any problems that may arise. Happy measuring!

Chapter 18: Understanding Trigger Jitter and Noise

Sources of Trigger Jitter and Noise

Digital oscilloscopes are powerful and versatile tools for measuring electronic signals. They allow engineers and technicians to analyze waveforms and troubleshoot complex circuits with precision and accuracy. However, when it comes to measuring fast and complex signals, trigger jitter and noise can significantly impact the quality and accuracy of measurements. In this chapter, we will explore the various sources of trigger jitter and noise and how they can affect your oscilloscope measurements. Trigger jitter is the variation in the time from the trigger signal to the acquisition of the waveform. This can make it challenging to trigger on signals with fast rise times or high frequency components. On the other hand, trigger noise is the random noise introduced by the trigger circuit, which can cause false triggering and measurement errors. Both trigger jitter and noise can be caused by internal and external factors, and it is essential to understand their sources to minimize their impact on measurements.

External Sources of Trigger Jitter and Noise

External signals, such as noise and interference, can significantly impact the trigger signal and introduce jitter and noise into your measurements. These external factors can be divided into two categories: environmental and signal-specific. Environmental factors include power line noise, electromagnetic interference (EMI), and ground loops. Power line noise is caused by the alternating current from power lines coupling into the oscilloscope's input circuitry. EMI is electrical noise generated by nearby devices, and ground loops are created by varying ground potentials between devices. To minimize the effects of these environmental factors, it is essential to have proper ground connections and use shielded cables.

Signal-specific factors include crosstalk, ringing, and transmission line effects. Crosstalk occurs when signals from one channel interfere with signals in adjacent channels, causing false triggering and measurement errors. Ringing is the high-frequency oscillation seen in square wave signals, which can introduce noise into the trigger signal. Transmission line effects occur when the length of the cables used in the measurement setup is not taken into account, causing impedance mismatches and reflections. To minimize the effects of these signal-specific factors, it is crucial to use proper grounding techniques and high-quality cables.

Internal Sources of Trigger Jitter and Noise

Although external sources of trigger jitter and noise can significantly impact measurements, internal factors within the oscilloscope can also contribute to these issues. Internal jitter and noise can occur due to timing errors in the trigger circuitry, signal processing, and sampling rate limitations. These factors can become more pronounced at higher frequencies and when using complex measurement setups. To minimize internal sources of trigger jitter and noise, it is crucial to select a high-quality oscilloscope with advanced trigger circuitry and high sampling rates. Additionally, regular calibration and maintenance of the oscilloscope can help reduce internal jitter and noise and ensure accurate measurements.

In summary, external and internal sources of trigger jitter and noise can significantly impact the accuracy and quality of oscilloscope measurements. As such, it is crucial to understand these sources and take the necessary measures to minimize their effects.

Techniques for Minimizing the Impact of Trigger Jitter and Noise

Now that we have explored the various sources of trigger jitter and noise, let's discuss some techniques for minimizing their impact on oscilloscope measurements. Here are a few tips to help you get more accurate and reliable measurements from your oscilloscope:

1. Proper Grounding and Shielding

As mentioned earlier, proper grounding and shielding can significantly reduce the effects of external sources of trigger jitter and noise. Make sure your oscilloscope is connected to a solid ground and use shielded cables to minimize interference.

2. Use Differential Probes

Differential probes can help eliminate noise and crosstalk by measuring the voltage difference between two points, rather than referencing to ground. This can provide more accurate and noise-free measurements, especially when dealing with high-frequency signals.

3. Adjust Trigger Levels and Slopes

Adjusting the trigger levels and slopes can also help minimize the effects of jitter and noise on measurements. By optimizing the trigger settings, you can ensure that the oscilloscope only captures the desired waveform and not any unwanted noise or jitter.

4. Use High-quality Cables and Connectors

Using high-quality cables and connectors is crucial, especially when working with high-frequency signals. Poor-quality cables can introduce impedance mismatches, reflections, and crosstalk, which can significantly impact measurement accuracy.

5. Keep Your Oscilloscope Clean and Well-maintained

Regular calibration and maintenance of your oscilloscope can also help minimize internal sources of trigger jitter and noise. Dust and dirt accumulation, as well as aging components, can affect the performance of the oscilloscope, leading to inaccuracies in measurements.

In conclusion, trigger jitter and noise must be taken into consideration when using an oscilloscope for measurements. By understanding the sources of these issues and implementing the techniques above, you can significantly improve the accuracy and reliability of your oscilloscope measurements.

Chapter 19: Further Troubleshooting with Oscilloscopes

Tips for Optimizing Your Oscilloscope Usage

Oscilloscopes are powerful tools that can help you troubleshoot and analyze electronic circuits. But in order to truly harness their potential, it is important to use them correctly. Here are some tips for optimizing your oscilloscope usage:

Choose the Right Probe

One of the most important factors in getting accurate measurements with your oscilloscope is using the correct probe for the application. There are various probes available, each designed for specific purposes such as high voltage measurements, high frequency signals, and differential signals. It is crucial to select the appropriate probe for the specific measurement you are trying to make. Using the wrong probe can lead to inaccurate results and even damage to the oscilloscope.

Properly Compensate Your Probe

Compensation refers to adjusting the probe to work correctly with the oscilloscope. A poorly compensated probe can result in distorted waveforms and incorrect measurements. Make sure to read the manufacturer's instructions for properly compensating your probe, and do it every time you change settings or switch to a new probe.

Use the Correct Settings and Probes for Probing High-Frequency Signals

Probing high-frequency signals can be a challenge, as the wires and the oscilloscope's input capacitance can affect and distort the signal. To minimize these effects, use the shortest possible connection between the signal and the oscilloscope. Also, use a probe with a higher bandwidth to mitigate signal distortions. Additionally, using a 10x attenuator can help properly load the signal on the oscilloscope, resulting in more accurate measurements.

Utilize the Waveform Math Functions

Many oscilloscopes come with built-in math functions that allow you to perform calculations on waveforms, such as addition, subtraction, multiplication, and more. These functions can be beneficial in troubleshooting complex circuits and understanding the relationships between different signals. Make use of these functions to save time and gain deeper insights into your circuits.

Understand and Use Triggering Effectively

Triggering is a crucial feature in oscilloscopes that allows you to capture a specific event or a specific part of a waveform. Using the correct trigger settings can help you capture elusive signals and make troubleshooting a lot easier. Familiarize yourself with the various trigger settings and how they affect the waveform display to use this feature effectively.

Take Advantage of Advanced Applications

Modern oscilloscopes come with advanced applications and features that can greatly improve your troubleshooting process. These applications include bus analysis, power integrity measurements, automated measurements, and more. Be sure to explore your oscilloscope's capabilities and utilize these applications when applicable to streamline your troubleshooting process.

Regularly Calibrate Your Oscilloscope

Calibrating your oscilloscope is crucial to ensure accurate and reliable measurements. It is recommended to calibrate your oscilloscope at least once a year, and more frequently if needed. Regular calibration ensures the accuracy and reliability of your measurements, leading to more efficient troubleshooting and analysis.

Stay Up-to-Date with Latest Technologies and Trends

Technology is constantly evolving, and so are oscilloscopes. Stay updated with the latest technologies and trends in oscilloscopes to make the most out of your troubleshooting process. Attend seminars, read articles, and network with other professionals in the field to stay on top of the latest developments.

In Conclusion

As you can see, there are many factors to consider when optimizing your oscilloscope usage for troubleshooting. From using the correct probe and settings to taking advantage of advanced applications, these tips can greatly improve your efficiency and accuracy in troubleshooting electronic circuits. Remember to regularly calibrate your oscilloscope and stay informed about the latest technologies to make the most out of your troubleshooting process. Happy oscilloscope troubleshooting!

Chapter 20: Common Applications of Oscilloscopes

Overview of Applications in Various Fields

Digital oscilloscopes are versatile tools that have revolutionized the way we observe and analyze electrical signals. With their ability to capture and display waveforms in real-time, digital oscilloscopes have become an indispensable instrument in a wide range of industries and fields. From electronics to telecommunications and medicine, these devices have found numerous applications that have greatly benefited professionals and researchers alike.

Electronics

In the field of electronics, oscilloscopes are used for a variety of purposes, such as circuit analysis, troubleshooting, and design verification. With their precise and accurate measurements, these devices allow engineers to visualize and analyze signals from various electronic components, such as resistors, capacitors, and inductors. By displaying waveforms on a screen, oscilloscopes provide valuable insights into the behavior of circuits and help in identifying issues and faults.

Moreover, oscilloscopes also play a crucial role in the design and development of electronic systems. With their advanced features, such as advanced triggering and automated measurements, engineers can quickly evaluate and validate their designs, ensuring their functionality and performance. Additionally, oscilloscopes with specialized probes and capabilities, such as power analysis and mixed-signal measurements, have become essential tools in fields such as power electronics, telecommunications, and automotive, where complex systems and high-frequency signals call for unique measurement techniques.

Telecommunications

In the ever-evolving world of telecommunications, oscilloscopes play a crucial role in ensuring the smooth and efficient transmission and reception of signals. With their fast sampling rates and high bandwidths, oscilloscopes are essential for the analysis of high-speed digital signals, such as those used in fiber optic communication, wireless networks, and satellite communication.

Moreover, oscilloscopes have also become integral in the development and testing of telecommunications equipment. From testing and characterizing radio frequency (RF) components to evaluating the quality of voice and data signals, these devices provide the necessary tools for engineers to measure and analyze critical parameters in communication systems. Additionally, oscilloscopes with specialized software and hardware options, such as eye diagrams and jitter analysis, allow engineers to delve deeper into the performance and reliability of their designs.

Medicine

The medical field is yet another area where oscilloscopes have found valuable applications. With the development of medical equipment and devices that utilize electrical signals, such as electrocardiograms (ECG), electromyograms (EMG), and electroencephalograms (EEG), oscilloscopes have become essential in the diagnosis and treatment of various medical conditions.

By capturing and displaying electrical signals from the human body, oscilloscopes provide doctors and medical professionals with critical insights into the functioning of the cardiovascular, muscular, and nervous systems. Moreover, with the advancement of technology and the emergence of telemedicine, oscilloscopes have also become vital in remote patient monitoring. With their ability to capture and store data, these devices provide doctors with a detailed history of a patient's condition, aiding in accurate diagnoses and treatment.

In Conclusion

With their vast capabilities and applicability in various fields, digital oscilloscopes have proven to be indispensable tools for professionals and researchers. From electronics to telecommunications and medicine, the versatility and precision of these devices have greatly advanced our understanding and manipulation of electrical signals. With constant advancements in technology, oscilloscopes are only bound to become more sophisticated and relevant in the future, leading to further progress and innovation in a wide range of industries and fields.

Chapter 21: Understanding Waveform Shapes and Patterns in Digital Oscilloscopes

Welcome back to our journey learning about digital oscilloscopes! In this chapter, we will dive deeper into the world of waveforms and how to interpret them on your oscilloscope. This is a fundamental skill for any oscilloscope user, as it allows you to understand the behavior of signals and diagnose any issues in your circuits.

The Basics of Understanding Waveform Shapes and Patterns

Before we can dive into identifying specific components and parameters in waveforms, we must first understand the basics of waveforms. A waveform is a graphical representation of how a signal changes over time. In an oscilloscope, waveforms are typically displayed on a Cartesian coordinate system, with voltage on the y-axis and time on the x-axis. The shape and pattern of a waveform can tell us a lot about the signal it represents. It can indicate the frequency, amplitude, and any distortions or abnormalities in the signal. By understanding the different types of waveforms and their characteristics, we can gain valuable information about the behavior of the signal and the circuit.

Identifying Specific Components and Parameters

Now that we have a basic understanding of waveform shapes and patterns, let's delve into how we can identify specific components and parameters in our signals. First and foremost, it is important to note that waveforms are not just random patterns on the screen. They follow specific patterns that correspond to the type of signal and the circuit components. The most common components seen in waveforms are the square, sine, triangular, and sawtooth waveforms. Each of these waveforms has distinct characteristics that allow us to identify them easily. For example, a square wave has a fast rise and fall time, while a sine wave has a smooth and rounded shape. Besides the shape of the waveform, other parameters that can be identified include frequency, amplitude, and phase. Frequency is the number of cycles of a waveform that occur in one second and is measured in Hertz (Hz). Amplitude is the measure of the height of a waveform, and it indicates the strength of the signal. The phase refers to the shift in time between two waveforms.

In order to accurately identify these components and parameters, it's important to use the measurement tools available on your oscilloscope. These tools allow you to measure and display these parameters, making it easier to understand the behavior of the signal.

Using Your Knowledge of Waveforms to Diagnose Issues

Now that we have a solid understanding of waveform shapes and patterns, we can use this knowledge to diagnose issues in our circuits. One of the most common uses for oscilloscopes is troubleshooting and debugging circuits. With a good understanding of waveforms, you can quickly pinpoint where the problem lies in your circuit. For example, if you notice a distorted or noisy waveform, it could be an indication of a malfunctioning component or a ground loop issue. By analyzing the shape and pattern of the waveform, you can narrow down the possible causes and take appropriate measures to fix the issue.

In addition to troubleshooting, understanding waveforms can also help with circuit design and optimization. By analyzing waveforms during circuit simulations, engineers can ensure the desired signal is being produced and make necessary adjustments to improve the circuit's performance.

Conclusion

Congratulations on learning about the basics of understanding waveform shapes and patterns on digital oscilloscopes! This knowledge will be a valuable asset in your journey to become an expert oscilloscope user. Remember to always pay close attention to the shape and pattern, as well as the parameters, of the waveforms on your oscilloscope. This will help you troubleshoot issues, optimize circuits, and gain a deeper understanding of the signals in your circuits. Keep practicing and experimenting, and you will become a waveform master in no time!

Chapter 22: Tips for Accurate Measurements

Digital oscilloscopes are powerful tools for analyzing electronic signals and diagnosing issues in circuits. However, in order to get accurate measurements, it is important to use the proper techniques and settings. In this chapter, we will discuss various techniques for getting the most accurate measurements with your digital oscilloscope.

Understanding Sampling Rates and Averaging

One of the key factors in getting accurate measurements with a digital oscilloscope is the sampling rate. The sampling rate is the number of data points collected per second and is typically measured in megasamples per second (MS/s). A higher sampling rate means more data points and therefore more accuracy in your measurements. Additionally, using averaging can also improve the accuracy of your measurements. Averaging takes multiple samples and calculates the average value, which can help smooth out any noise in the signal. This is particularly useful when measuring signals with high frequency or inconsistent signals.

Set Your Trigger Correctly

Having the correct trigger settings is crucial for accurate measurements. The trigger is the signal that tells the oscilloscope when to start capturing data. If the trigger is not set correctly, the captured waveform may not be an accurate representation of the signal. Make sure to set the trigger to the correct threshold level and trigger mode for your signal. This will ensure that the oscilloscope is triggered at the appropriate point in the waveform.

Use Correct Probe Settings

Choosing the right probe and compensating it properly can also greatly impact the accuracy of your measurements. Using a probe with too high of an impedance can cause reflections and distort the waveform, while a low impedance probe can load the circuit and also cause inaccuracies.

Probes should also be compensated, which means adjusting it to match the input impedance of the oscilloscope. This ensures that the probe does not affect the voltage and frequency of the signal being measured.

Minimize Noise and Interference

Inaccuracies in measurements can also be caused by external noise and interference. To minimize these effects, make sure to properly ground your oscilloscope and use shielded cables when necessary. You can also use the oscilloscope's built-in filters to reduce noise and interference. Additionally, use proper insulation and avoid placing the oscilloscope near sources of interference, such as other electronic devices or power cables.

Adjust Time and Voltage Scales

To get the most accurate representation of your waveform, it is important to adjust the time and voltage scales accordingly. This means adjusting the horizontal time scale to capture multiple cycles of the signal and the vertical voltage scale to ensure the entire waveform is visible on the screen. Using autorange can often lead to less accurate measurements, so it is better to manually adjust the scales to fit the signal.

Calibrate Your Oscilloscope

Regular calibration is essential for maintaining the accuracy of your oscilloscope. Most oscilloscopes have a self-calibration feature that can be performed regularly to ensure the instrument is functioning properly. Additionally, it is recommended to perform a full calibration at least once a year. This calibration can be done by a professional calibration service or by using a calibration kit specifically designed for oscilloscopes.

Conclusion

Accurate measurements are vital for accurately diagnosing and troubleshooting electronic circuits. By following these tips and techniques, you can ensure that your digital oscilloscope is providing you with the most accurate measurements possible. Remember to adjust your settings, use proper probes, and minimize noise and interference for the best results. And don't forget to regularly calibrate your instrument for continued accuracy.

Chapter 23: Reducing Noise and Interference in Oscilloscope Measurements

Strategies for Reducing Noise and Interference

As any oscilloscope user knows, obtaining accurate and reliable measurements can be a challenging task. With the ever-increasing complexity of modern electronic circuits, the chances of introducing noise and interference into your measurements are higher than ever before. However, this does not mean that all hope is lost. With the right techniques and strategies, you can significantly reduce the effects of noise and interference on your measurements and obtain precise results.

Proper Grounding Techniques

One of the key strategies for reducing noise and interference in oscilloscope measurements is proper grounding. Grounding is the process of connecting the ground terminal of your oscilloscope probe to a reference point in a circuit. This reference point is usually the ground or earth connection of the circuit. By doing so, you are creating a stable and low-impedance path for the noise and interference to flow away from the measured signal. When grounding your oscilloscope, it is crucial to follow proper techniques to ensure the best results. One common mistake that many users make is grounding their probe to the wrong reference point. It is essential to identify the true ground of your circuit before connecting your probe. If you ground to a false or floating ground, you may introduce more noise and interference into your measurements, resulting in inaccurate results.

Another key aspect of proper grounding is to use short, low-impedance ground leads. The longer the ground lead, the higher the chances of picking up noise and interference. Therefore, selecting a ground lead that is as short and direct as possible can significantly reduce the effects of external interference.

Shielding and Filtering Techniques

Shielding and filtering are two strategies that use physical barriers and components to minimize the impact of noise and interference. Shielding involves covering your oscilloscope and probe with a metallic shield to block out external interference sources such as mobile phones, radios, and other electronic devices. If you are working in an area with high levels of electromagnetic interference, it is essential to use a shielded oscilloscope to obtain accurate measurements.

Filtering, on the other hand, involves using electronic components such as capacitors, inductors, and resistors to remove unwanted noise and interference from the measured signal. In some cases, you may need to use a combination of both shielding and filtering techniques to achieve optimal results.

Signal Routing and Probing Techniques

The way you route your signals and probe your circuit can also play a significant role in reducing noise and interference in your measurements. One common mistake that many oscilloscope users make is routing the signal and ground leads too close to each other, resulting in cross-talk. Cross-talk occurs when the ground lead interferes with the measured signal, leading to distorted results. To prevent cross-talk, it is crucial to keep the signal and ground leads separated as much as possible. If you are probing a high-frequency signal, make sure to use a short, direct path for the signal lead and a separate ground lead. For lower frequency signals, you can twist the signal and ground leads together to reduce the effects of cross-talk.

Moreover, when probing, try to avoid touching the probe tip with your fingers or any other objects. Doing so can create additional noise and interference, affecting the accuracy of your measurements.

Ground Loops and Differential Probing

Another source of noise and interference in oscilloscope measurements is ground loops. Ground loops occur when there is more than one path for current to flow between ground points. This can create noise and interfere with the measured signal.

To avoid ground loops, it is crucial to use differential probing techniques whenever possible. Differential probing involves using two probes simultaneously, with one probe measuring the signal and the other measuring the ground. By doing so, you eliminate the possibility of ground loops and obtain more accurate measurements.

Conclusion

In conclusion, reducing noise and interference in oscilloscope measurements requires a combination of proper techniques and strategies. By following the guidelines mentioned above, you can significantly improve the accuracy and reliability of your measurements. Remember to always use proper grounding techniques, shield and filter your signals, carefully route and probe your circuits, and avoid ground loops to obtain the best results. With these strategies in mind, you can confidently make precise measurements and troubleshoot any issues with your digital oscilloscope.

Chapter 24: Troubleshooting Digital Circuits

Techniques for Troubleshooting Digital Circuits

As technology continues to advance at an astonishing pace, digital circuits have become an integral part of our daily lives. From smartphones, computers, and home appliances to complex industrial systems, digital circuits are everywhere. With the increasing complexity and sophistication of these circuits, it has become more challenging to identify and fix issues when they arise. In this chapter, we will explore some techniques for troubleshooting digital circuits, including identifying faulty logic levels and timing issues. These techniques will help you accurately diagnose and resolve any issues that may arise with your digital circuits.

Identifying Faulty Logic Levels

One of the most common issues faced when troubleshooting digital circuits is incorrect logic levels. This can lead to faulty operation, system crashes, or even hardware damage. So how do we identify and correct these issues?

Firstly, it is essential to understand the logic levels of the digital signals in your circuit. Most digital circuits operate on binary signals with two possible states - high (logic 1) and low (logic 0). However, the exact voltage levels for these logic states may vary depending on the circuit's specifications.

To identify faulty logic levels, you can use your oscilloscope to measure the voltage at specific points in the circuit. Compare these measurements to the expected voltage levels, and if they do not match, you may have identified a faulty logic level. You can then troubleshoot further by tracing the signal path and checking for any damaged components or faulty connections.

Timing Issues in Digital Circuits

In addition to checking for faulty logic levels, troubleshooting digital circuits also involves identifying timing issues. In a digital circuit, the timing of signals is crucial, and any deviations can lead to circuit malfunctions. One way to identify timing issues is to use your oscilloscope's time-based measurements and triggering capabilities. You can set up your scope to trigger on specific edge or pulse widths, allowing you to analyze the timing of signals at various points in the circuit. This can help you identify any timing issues and pinpoint where they are occurring. Another useful tool for troubleshooting timing issues is the use of logic analyzers. These instruments can analyze and decode digital signals, making it easier to spot any timing errors. They also offer advanced triggering and measurement features specific to digital circuits, making them an invaluable tool for troubleshooting.

Other Tips for Troubleshooting Digital Circuits

In addition to the techniques mentioned above, here are a few other tips that can help you troubleshoot digital circuits more efficiently:

- Check for loose connections or damaged components: Sometimes, the issue may be as simple as a loose connection or a damaged component. Always double-check all connections and visually inspect your circuit for any obvious signs of damage.
- Use signal integrity measurements: Modern oscilloscopes offer advanced features to measure and analyze signal integrity, such as duty cycle, rise and fall times, and overshoot and undershoot. These can help you detect any issues with the quality of your digital signals.
- Verify your waveform math: Many oscilloscopes allow you to perform simple math functions on waveforms, such as addition, subtraction, and multiplication. If you are troubleshooting a complex digital circuit, make sure to verify your math functions to rule out any errors in your results.
- Use logic probes: Logic probes are handheld instruments used to measure digital signals. They can be handy when tracing signals through a complex circuit, as they provide visual indicators for logic levels.
- Refer to the circuit diagram: Lastly, always refer to the circuit diagram of your digital circuit while troubleshooting. It can help you understand the signal path and pinpoint where the issue may be occurring.

In Conclusion

Troubleshooting digital circuits can be a challenging task, but with the right techniques and tools, it can be done efficiently and accurately. By understanding the logic levels and timing requirements of your circuit, along with utilizing advanced oscilloscope features and logic analyzers, you can quickly identify and resolve any issues that may arise. Don't forget to also use our tips for troubleshooting digital circuits to make your troubleshooting process more efficient. Remember, practice makes perfect, so keep honing your skills, and soon you will become a pro at troubleshooting digital circuits.

Chapter 25: Advanced Measurements with Mixed-Signal Capability

Overview of Mixed-Signal Measurement Capabilities

Oscilloscopes have come a long way from their humble origins as a tool for displaying electrical signals on a screen. With the technological advancements in digital electronics, modern oscilloscopes also have the capability to measure and analyze both analog and digital signals, making them an essential tool for debugging complex systems.

The ability to measure both analog and digital signals simultaneously is known as mixed-signal measurement capability. This feature has become increasingly important as more and more systems involve a combination of analog and digital components.

How to Set Up and Interpret Results

Setting up your oscilloscope for mixed-signal measurements may seem daunting at first, but it is actually quite straightforward. The key components you will need are a mixed-signal probe and a waveform generator. The mixed-signal probe is essentially two probes in one; it has a ground clip and two signal clips, allowing you to connect to both analog and digital signals simultaneously. To set up your oscilloscope for mixed-signal measurements, start by connecting the mixed-signal probe to channel 1 of your oscilloscope. Then, connect the analog signal to the positive input and the digital signal to the negative input of the probe. This will allow you to measure both signals on the same display.

Next, connect your waveform generator to channel 2 of your oscilloscope and set it to generate a square wave. This will serve as your digital signal for testing and analysis. Once you have both signals connected, you can adjust the settings on your oscilloscope to accurately measure and analyze your signals. You can adjust the vertical and horizontal scales, trigger settings, and measurement parameters to get the most accurate results. It is also important to note that when working with mixed-signal measurements, you may need to adjust the input impedance of your oscilloscope. Most oscilloscopes have a default input impedance of 1 MΩ and 10 pF, but for mixed-signal measurements, it is recommended to adjust the impedance to 1 MΩ and 15 pF. This will help minimize any noise or interference between the two signals. Once you have your signals set up and your oscilloscope properly adjusted, you can start interpreting your results. For example, you can use the oscilloscope's cursors to measure the rise and fall times of your digital signal and compare them to the specifications of your waveform generator.

You can also use the oscilloscope's built-in math functions to perform calculations on both signals, providing valuable insights into the behavior of your system. In addition to basic measurements, advanced oscilloscope features also allow for more in-depth analysis of mixed-signal systems. For example, the oscilloscope's serial bus decoding capability allows you to capture and decode digital signals in various protocols, such as I2C, SPI, and UART. This is especially useful in debugging communication protocols between analog and digital components. Moreover, modern oscilloscopes also offer features such as digital phosphor and color grading, which help clearly differentiate between analog and digital signals on the display. This makes it easier to identify any potential timing or signal integrity issues between the two components.

In conclusion, the mixed-signal measurement capabilities of modern oscilloscopes make them an indispensable tool for debugging complex systems. By following the steps outlined in this chapter and making use of advanced features, you can accurately measure and analyze both analog and digital signals for a deeper understanding of your system's behavior. So don't shy away from mixed-signal measurements, embrace them with confidence and make the most out of your oscilloscope's capabilities.

Chapter 26: Power Analysis with Oscilloscopes

Using Oscilloscopes for Power Measurements

When it comes to electronics, understanding power is crucial. In order to design, troubleshoot, and optimize circuits, it is important to have accurate measurements of power consumption. Oscilloscopes are powerful tools that can provide valuable insight into power analysis. With their ability to capture waveforms in real-time, oscilloscopes can accurately measure AC, DC, and mixed-signal power signals.

AC and DC Power Analysis

The two main types of power signals are AC (alternating current) and DC (direct current). AC signals change over time, whereas DC signals have a constant value. Oscilloscopes are capable of measuring both types of signals and can display the power waveform in real-time. By using the AC and DC coupling modes, the oscilloscope can accurately measure the voltage and current waveforms separately, allowing for more detailed power analysis. When measuring AC power, it is important to consider the frequency of the signal. High frequency AC signals can easily distort the waveform, making accurate measurements a challenge. However, modern oscilloscopes have advanced features such as bandwidth limiting and low-noise front-end amplifiers that can help mitigate these issues.

DC power analysis, on the other hand, allows for more static measurements of power consumption. With the ability to capture and display stable waveforms, oscilloscopes can accurately measure DC voltage and current levels. This is especially useful when analyzing the overall efficiency of a circuit, as it provides a snapshot of the power consumption at a specific moment in time.

Power Supply Troubleshooting

Power supply issues are a common problem in electronic circuits. Whether it's due to faulty components or improper design, a malfunctioning power supply can cause issues such as overheating, damage to components, and overall poor performance. This is where oscilloscopes can be a valuable tool for troubleshooting. By using an oscilloscope to measure the voltage and current waveforms of the power supply, engineers can easily spot any abnormalities or fluctuations. These can include ripple voltage, noise, and voltage spikes, all of which can affect the overall performance of the circuit. With advanced trigger and measurement options, oscilloscopes can help isolate specific components or areas of the circuit that may be causing the issue.

Furthermore, oscilloscopes can also be used to analyze the efficiency of a power supply. By measuring the input and output power, engineers can calculate the overall efficiency of the supply. This can help identify any areas that need improvement, such as reducing power losses and improving overall performance.

Conclusion

In conclusion, oscilloscopes are powerful tools for power analysis in electronics. Their ability to capture and display real-time waveforms allows for accurate measurements of AC, DC, and mixed-signal power signals. Whether it's for measuring power consumption, troubleshooting power supply issues, or analyzing the efficiency of a circuit, oscilloscopes provide valuable insights that can help improve overall performance and functionality. With their advanced features and capabilities, oscilloscopes continue to play a vital role in the world of electronics and power analysis.

Chapter 27: Advanced Applications of Oscilloscopes

Comparing and Contrasting the Functionality of Oscilloscopes and Spectrum Analyzers for Frequency Domain Analysis

Oscilloscopes and spectrum analyzers are two powerful tools used for frequency domain analysis in various industries such as electronics, telecommunications, and aerospace. While they both serve the same purpose of analyzing signals in the frequency domain, they have distinct differences in their functionality and capabilities. In this chapter, we will explore the similarities and differences between oscilloscopes and spectrum analyzers and how they can be used for advanced applications.

Understanding the Basics of Frequency Domain Analysis

Before diving into the comparison between oscilloscopes and spectrum analyzers, it is important to understand the basics of frequency domain analysis. In simple terms, frequency domain analysis involves analyzing the frequency components of a signal. This is done by converting a time-domain signal into its frequency domain representation using a mathematical process called Fourier analysis. This allows for a deeper understanding of a signal and its characteristics, such as amplitude, frequency, and phase.

The Functionality of Oscilloscopes for Frequency Domain Analysis

Oscilloscopes are versatile measurement instruments that are commonly used for time-domain analysis of signals. However, they also have the ability to perform frequency domain analysis through the use of a Fast Fourier Transform (FFT). This feature allows an oscilloscope to display the frequency components of a signal in a spectrum analyzer-like format.

One of the main advantages of using an oscilloscope for frequency domain analysis is its wide bandwidth. Oscilloscopes are known for their high bandwidth capabilities, which allows for accurate analysis of high-frequency signals. They also have a high sampling rate, which is essential for capturing transient signals accurately. However, the frequency resolution of oscilloscopes may not be as precise as that of a spectrum analyzer, making them better suited for analyzing signals with varying frequencies rather than a specific frequency.

The Functionality of Spectrum Analyzers for Frequency Domain Analysis

Spectrum analyzers, on the other hand, are specifically designed for frequency domain analysis. They offer high frequency resolution and a wide dynamic range, making them ideal for analyzing signals with a specific frequency. Unlike oscilloscopes, spectrum analyzers do not capture signals in the time domain but rather analyze and display the frequency components of a signal directly. Spectrum analyzers also have the ability to perform in-depth analysis of signals in the frequency domain, such as measuring power, harmonic distortion, and modulation. This makes them a valuable tool for troubleshooting and characterizing various signals in a given frequency range. However, spectrum analyzers have limited bandwidth compared to oscilloscopes, making them less suitable for analyzing high-frequency signals.

The Advantages and Disadvantages of Using Oscilloscopes and Spectrum Analyzers

The choice between using an oscilloscope or spectrum analyzer for frequency domain analysis ultimately depends on the specific application and requirements. Both instruments have their own advantages and limitations, and it is important to understand them in order to determine the most suitable tool for a given task. For applications that require accurate analysis of high-frequency signals, oscilloscopes are the preferred option due to their high bandwidth and sampling rate. On the other hand, spectrum analyzers excel in analyzing signals with a specific frequency and providing in-depth analysis in the frequency domain. However, spectrum analyzers may not be as useful for analyzing signals with varying frequencies.

In Conclusion

In conclusion, oscilloscopes and spectrum analyzers serve different purposes in the realm of frequency domain analysis. While both instruments offer similar functionalities, they have distinct advantages and limitations that make them better suited for different applications. By understanding the differences between oscilloscopes and spectrum analyzers, engineers and technicians can choose the right tool for their specific needs and achieve accurate and reliable results.

Chapter 28: Debugging, Decoding & Analysis

Oscilloscopes are powerful tools for measuring and analyzing electronic signals. While they are commonly used for basic measurements and troubleshooting, they also have advanced features that make them valuable for more complex applications. In this chapter, we will explore some of the advanced features of oscilloscopes, including advanced triggering options and bus decoding capabilities. These features allow for more efficient debugging and analysis, making oscilloscopes a must-have tool for any electronics engineer.

Advanced Features for Efficient Debugging

In order to efficiently debug complex circuits, engineers need tools that can accurately capture and analyze signals. Oscilloscopes have a range of advanced features that make it easier to troubleshoot and identify issues in electronic systems. These features include advanced triggering options, which allow for more precise and accurate triggering of signals, and bus decoding capabilities, which enable engineers to analyze digital signals transmitted over various communication protocols. One of the most useful advanced triggering options is the ability to set up multiple triggers at once. This feature, known as "smart triggers," allows engineers to set up triggers based on multiple signal parameters such as amplitude, slew rate, or pulse width. With smart triggers, engineers can set up precise trigger conditions for complex signals, making it easier to identify and capture intermittent signal abnormalities.

Another useful triggering option is the ability to use logic-based triggers. This feature allows users to trigger on logic-level signals, making it ideal for debugging digital circuits. Engineers can use this feature to trigger on a specific bit or bit pattern, which is excellent for identifying specific problems in digital systems. Logic-level triggers also allow for triggering on both rising and falling edges of the signal, providing a more comprehensive view of the signal.

Bus Decoding Capabilities

In modern electronic systems, signals are often transmitted over various communication protocols such as I2C, SPI, UART, and CAN. Analyzing these protocols can be a daunting task for an engineer without the proper tools. However, oscilloscopes with bus decoding capabilities make it much easier to decode and analyze digital signals transmitted through these protocols.

With bus decoding, engineers can see decoded bus signals overlaid on the waveform, making it easier to identify and troubleshoot problems in communication between devices. Additionally, oscilloscopes with bus decoding capabilities can also display bus statistics, such as error rates, making it easier to pinpoint issues in the communication protocol.

Real-time Serial Protocol Analysis

Advanced oscilloscopes also have real-time serial protocol analysis capabilities, allowing for a comprehensive overview of digital signals. With real-time protocol analysis, engineers can analyze protocols while the system is running, capturing important data and providing insights into the system's behavior. This feature is particularly useful when trying to identify root causes of communication failures or protocol errors. Additionally, oscilloscopes with this capability can also trigger on specific protocol events and display decoded packets and errors in real-time. This level of analysis is invaluable when debugging complex systems or identifying issues in communication between devices.

Enhanced Measurement Capabilities

Another way advanced oscilloscopes aid in efficient debugging is through enhanced measurement capabilities. These capabilities include advanced math functions, such as FFT analysis, advanced measurements for signals with fast rise times, and power analysis features. FFT analysis, or Fast Fourier Transform, is an advanced math function that allows engineers to analyze complex waveforms and identify frequency components. This feature is especially useful when troubleshooting noise and interference in signals. Engineers can use FFT analysis to determine the frequency of unwanted signals and take steps to reduce or eliminate them. Advanced measurements for signals with fast rise times are also essential for efficient debugging.

Oscilloscopes with this feature can accurately measure signals with fast rise times, which are often missed by traditional measurement methods. This capability is particularly useful when testing high-speed digital circuits, where signals can change rapidly. Additionally, advanced oscilloscopes can also perform power analysis, which is crucial for measuring the performance of low-power electronic devices and systems. With power analysis features, engineers can analyze power usage, potential sources of power losses, and identify areas for improvement.

Conclusion

In conclusion, advanced oscilloscopes are essential tools for any engineer working with complex electronic systems. With advanced triggering options, bus decoding capabilities, real-time serial protocol analysis, and enhanced measurement capabilities, these tools make it easier to identify, troubleshoot, and resolve issues in electronic systems. As technology continues to advance, so do oscilloscope capabilities, making them irreplaceable in the world of electronics engineering. Keep exploring the vast range of advanced features offered by oscilloscopes and unlock their full potential in your projects. Happy debugging!

Chapter 29: Advanced Applications of Oscilloscopes

Techniques for reducing noise and interference in power measurements

When it comes to making accurate power measurements, noise and interference can greatly affect the results. As we have discussed, power measurements can be a tricky task as they involve high frequency signals and fast rise times. These characteristics make them susceptible to noise and interference, leading to distorted waveforms and incorrect measurements. However, with the right techniques, we can greatly reduce the impact of noise and interference on our power measurements, ensuring accurate and reliable results.

Using current probes

One of the most commonly used techniques for reducing noise and interference in power measurements is by using current probes. These probes are specifically designed to measure current passing through a conductor and have a high bandwidth and sensitivity, making them ideal for measuring high frequency signals. By using a current probe, we can isolate the signal of interest from any unwanted noise or interference picked up along the way.

When using a current probe, it is essential to properly calibrate and compensate for accurate readings. This involves adjusting the probe's gain and compensation settings to match the oscilloscope's input impedance. By doing so, we can reduce any added noise or distortion caused by the current probe itself. Properly compensating the current probe is crucial, especially when working with fast rise time signals. A poorly compensated probe can introduce reflections and overshoots, resulting in inaccurate measurements.

Differential probes

Another effective technique for reducing noise and interference in power measurements is by using differential probes. These probes are designed to measure the voltage difference between two points, eliminating any common-mode noise or interference. This can be especially useful when working with high frequency signals where common-mode noise can significantly affect the accuracy of our measurements. Differential probes have a high common-mode rejection ratio, allowing us to focus on the signal of interest while rejecting any unwanted noise or interference.

When using a differential probe, it is crucial to correctly connect it to the oscilloscope. The channel A and B inputs on the oscilloscope should be connected to the inputs of the probe, while the output of the probe should be connected to the single-ended input channel on the oscilloscope. It is also essential to ensure that the probe's ground lead is connected to the proper ground reference for accurate measurements.

Proper grounding techniques

Along with using appropriate probes, proper grounding techniques also play a vital role in reducing noise and interference in power measurements. Ground loops can be a significant source of interference, as they create unwanted current paths that can influence our measurements. By using proper grounding techniques, we can minimize the impact of ground loops and ensure accurate measurements. One important technique for proper grounding is to use a star grounding scheme. This involves using a single point as the central ground reference and connecting all the grounds from various sources to this point. By doing so, we can reduce the chances of ground loops and minimize any unwanted interference. Additionally, keeping ground leads as short as possible can also help reduce the impact of ground loops.

Proper shielding techniques

In addition to proper grounding, using shielding techniques can also help reduce noise and interference in power measurements. Shielded cables and enclosures can provide a barrier between the measured signals and any external noise sources, reducing the impact of interference on our measurements. It is essential to use proper shielding techniques when working with high frequency signals to maintain signal integrity.

When using shielded cables, it is crucial to connect the shield to a proper ground. A floating shield can act as a receiving antenna, picking up external interference and affecting our measurements. By connecting the shield to a proper ground, we can divert any external noise away from the measured signal, ensuring accurate results.

Isolation techniques

In some cases, external noise and interference can be too severe and affect our measurements, even after using proper grounding and shielding techniques. In such scenarios, using isolation techniques can help eliminate the interference and ensure accurate measurements. Isolation transformers and optocouplers are two common isolation techniques used in oscilloscopes. An isolation transformer can help break the ground loop by isolating the ground reference from the rest of the circuit. This can greatly reduce the impact of external interference on our measurements. Optocouplers, on the other hand, use light to transfer the signal, thus providing complete electrical isolation between the oscilloscope and the circuit under test. Using isolation techniques can be an effective solution when working with highly sensitive circuits that require a high level of accuracy.

Keeping the signal path short

One final technique for reducing noise and interference in power measurements is by keeping the signal path as short as possible. Long signal paths can act as antennas, picking up external interference and affecting our readings. By keeping the signal path short, we can minimize the impact of any interference and achieve more accurate measurements.

When working with high frequency signals, it is essential to use proper transmission line techniques to maintain a consistent impedance. This can help reduce reflections and attenuate any unwanted signals picked up along the way. Additionally, avoiding the use of long probe leads can also contribute to keeping the signal path short, reducing the impact of external noise and interference.

In Conclusion

Accurate power measurements are vital in many applications, and noise and interference can significantly affect their results. By using appropriate techniques, such as current probes and differential probes, along with proper grounding, shielding, and isolation techniques, we can greatly reduce the impact of noise and interference on our measurements. Additionally, keeping the signal path short and using proper transmission line techniques can also contribute to achieving more accurate and reliable power measurements. By utilizing these techniques, we can ensure the integrity of our measurements and obtain valuable insights into our circuits' performance.

Chapter 30: Tips for Probing High-Frequency Signals with Precision and Accuracy

Regardless of the complexity of a signal, oscilloscopes are powerful tools for measuring and analyzing electrical signals. However, when working with high-frequency signals, it is crucial to ensure the accuracy and precision of the measurements. Any slight disturbance or interference can significantly affect the signal quality and make it challenging to obtain accurate results. Therefore, in this chapter, we will discuss some essential tips for probing high-frequency signals without compromising signal quality, such as using active probes.

Tips for Probing High-Frequency Signals without Affecting Signal Quality

When working with high-frequency signals, the following tips can help you obtain precise and accurate measurements without affecting the signal quality:

Use Active Probes

Active probes are high-performance probes designed specifically for high-frequency signals. They offer a wider frequency range, lower loading effects, and better impedance matching compared to passive probes. The active circuitry of these probes reduces the capacitive loading effect on the circuit under test, resulting in more accurate measurements. Additionally, they have a higher sensitivity and lower noise levels, making them ideal for probing small signals without distortion.

Use Short Ground Leads

The longer the ground lead, the higher the inductance, which can cause unwanted reflections and distortions in the signal. To minimize the inductance, it is essential to use the shortest possible ground lead, preferably less than 10 cm. This will ensure a more direct connection to the ground, reducing any interference or parasitic capacitance that can affect the signal.

Minimize Wiring and Connections

Every connection and wiring adds capacitance and inductance, which can significantly impact high-frequency signals. Therefore, it is crucial to keep the circuit as simple as possible and minimize the number of connections. This will reduce any additional loading effects and minimize the risk of unwanted reflections and distortions in the signal.

Pay Attention to the Probe Tip

The tip of the probe is the main point of contact with the circuit under test. Therefore, it is essential to use a high-quality probe with a sharp and clean tip. A dull or dirty tip can cause unwanted reflections and distortions in the signal. It is also essential to utilize the appropriate probe tip size depending on the circuit's size and components. Using a smaller probe tip can result in a lower loading effect and better signal fidelity.

Use the Appropriate Attenuation Setting

When probing high-frequency signals, it is vital to use the appropriate attenuation setting to avoid overloading the oscilloscope's front end. If the signal is too strong, it can cause distortion and affect the accuracy of the measurements. Therefore, it is crucial to use an attenuation setting that can handle the amplitude of the signal without saturating the input.

Use Differential Probes for Differential Signals

Differential signals are commonly used in high-speed digital circuits to minimize noise and interference. Therefore, when probing these signals, it is essential to use a differential probe rather than two single-ended probes. Differential probes offer higher common-mode rejection, better bandwidth, and lower noise levels, resulting in more accurate measurements of differential signals.

Keep the Probes Close to the Device Under Test

When probing high-frequency signals, it is vital to keep the probes as close to the device under test as possible. This reduces the effects of inductance and capacitance, resulting from long probe leads and connections. The shorter the distance, the more accurate the measurements will be.

Conclusion

Probing high-frequency signals with precision and accuracy can be challenging, but by following the tips mentioned above, you can achieve reliable and accurate measurements. Active probes, short ground leads, minimal wiring and connections, paying attention to the probe tip, using the appropriate attenuation setting, using differential probes for differential signals, and keeping the probes close to the device under test are all crucial factors in obtaining accurate and distortion-free measurements. With the right techniques and tools, you can confidently work with high-frequency signals and troubleshoot any issues with ease.

Chapter 31: Techniques for Measuring Signals with Fast Rise Times

Introduction

Have you ever encountered a signal with a fast rise time that seems impossible to accurately measure? As technology continues to advance, signals with fast rise times are becoming more common in our daily lives. From high-speed digital signals in computers to high-frequency radio signals, being able to accurately capture signal details is crucial for proper troubleshooting and analysis. This chapter will explore different techniques for measuring signals with fast rise times, allowing you to confidently handle any signal that comes your way.

The Importance of Accurate Measurements

In today's fast-paced technological world, the importance of accurate measurements cannot be emphasized enough. A single faulty measurement can lead to incorrect analysis and troubleshooting, which can result in costly errors and delays. This is especially true when dealing with signals that have a fast rise time. These signals are extremely sensitive and can easily be distorted if not handled properly. Therefore, it is essential to have a solid understanding of different techniques for measuring these signals accurately.

Choosing the Right Oscilloscope

Before diving into the techniques for measuring signals with fast rise times, it is crucial to have the right tools. When it comes to fast signals, having an oscilloscope with a high bandwidth and sampling rate is essential. The bandwidth of an oscilloscope refers to its ability to accurately capture high-frequency signals, while the sampling rate is the number of samples per second the oscilloscope can capture. A high bandwidth and sampling rate allow the oscilloscope to capture rapid voltage changes accurately.

Using a Low-Inductance Probe

One of the first techniques to consider when measuring signals with fast rise times is using a low-inductance probe. Traditional oscilloscope probes have a high inductance, which means they can pick up noise and alter the signal being measured. This is especially true for signals with fast rise times. A low-inductance probe has a shorter ground lead, reducing the inductance and minimizing noise interference. This allows for a more accurate measurement of the signal's rise time.

Adjusting the Attenuation Factor

Another technique for measuring fast signals is adjusting the attenuation factor. Attenuation is a process that reduces the amplitude of a signal. Oftentimes, fast signals have a higher amplitude than what the oscilloscope can handle. By adjusting the attenuation factor, you can reduce the signal's amplitude and accurately measure its rise time. This technique is particularly useful when working with high-voltage signals.

Using a High-Speed Trigger

Many oscilloscopes come equipped with a high-speed trigger feature. This is specifically designed for measuring fast signals and helps to ensure accurate measurements. The high-speed trigger allows you to trigger the oscilloscope at a specific voltage level, helping to capture the signal's rise time without any distortion. It is essential to set the trigger level slightly higher than the expected signal amplitude to ensure it triggers at the desired time.

Enabling the Averaging Feature

For signals with extreme fluctuations and fast rise times, enabling the averaging feature can help improve the accuracy of your measurements. This feature can reduce noise and provide a more stable and accurate representation of the signal. However, it is crucial to keep in mind that using the averaging feature can increase the measurement time, so it should be used only when necessary.

Adjusting the Oscilloscope's Ground Level

In some cases, the distortion of a fast signal can be caused by ground level differences between the oscilloscope and the circuit being measured. To prevent this, adjust the oscilloscope's ground level to match the circuit's ground level. This will help ensure that the signal is measured accurately and reduce any potential distortion.

Using Math Functions for Signal Averaging

Most modern oscilloscopes come equipped with various math functions to aid in signal analysis. One of these functions is signal averaging, which can help reduce noise and provide a more accurate representation of the signal's rise time. By averaging multiple waveforms, you can get a clearer picture of the signal and its rise time.

Conclusion

Measuring signals with fast rise times can be a challenging task, but with the right techniques and tools, it can be done with precision and accuracy. By choosing the right oscilloscope, using a low-inductance probe, adjusting the attenuation, enabling the high-speed trigger, and utilizing various math functions, you can confidently measure any signal that comes your way. Remember, accurate measurements are the foundation for accurate analysis and troubleshooting, so it is crucial to have a solid understanding of techniques for measuring fast signals.

Chapter 32: Advanced Troubleshooting Techniques with Waveform Math Functions

Application of Waveform Math Functions to Analyze Complex Signals and Troubleshoot Issues

Digital oscilloscopes have revolutionized the way electronic signals are analyzed and tested. With the advancements in technology, the complexity of electronic circuits has increased exponentially. As a result, traditional methods of signal analysis and troubleshooting have become inadequate. This is where waveform math functions come into play. These powerful functions allow users to perform advanced calculations on acquired waveforms, providing valuable insight into the behavior of complex signals. In this chapter, we will explore the applications of waveform math functions and how they can be used to effectively troubleshoot issues in electronic circuits.

Understanding Waveform Math Functions

Waveform math functions, also known as math operations, are built-in tools in digital oscilloscopes that allow users to perform mathematical operations on acquired waveforms. These operations can be simple addition, subtraction, multiplication, or division, or they can be more complex functions like FFT, differentiation, and integration. These functions can be applied to one or more waveforms, providing a new waveform as a result. This new waveform can then be displayed on the oscilloscope screen, allowing users to visualize the effects of the mathematical operation on the original signal.

The Types of Waveform Math Functions and Their Applications

There are many types of waveform math functions available in digital oscilloscopes, each with its unique purpose and application. Let's take a look at some of the most common ones and how they can be used for troubleshooting complex issues.

FFT (Fast Fourier Transform)

FFT is a linear transformation that converts a time-domain waveform into a frequency-domain spectrum. This function is useful when analyzing signals with multiple frequency components or when identifying the frequency of a particular signal. It can also help identify sources of noise and interference in a signal. By using the FFT function, users can easily detect unwanted frequency components and troubleshoot issues that arise from them.

Differentiation and Integration

The differentiation function allows users to calculate the derivative of a waveform, while the integration function calculates the integral. These functions can be used to analyze the rate of change of a signal and its overall energy consumption. They are particularly useful in power analysis applications, where understanding the power consumption of a circuit is crucial. Differentiation and integration functions can also be used to identify noise and glitches, making them valuable tools for troubleshooting complex issues.

Advanced Triggering and Mask Testing

Waveform math functions can be used in conjunction with advanced triggering modes to identify anomalous signals and trigger acquisitions at specific events. This can be particularly useful when troubleshooting issues such as intermittent glitches and noise. By setting up a math function to detect certain characteristics of a signal and triggering on it, users can capture the exact waveform that caused the issue, making troubleshooting more efficient. Additionally, math functions can be used for mask testing, where predefined limits are set to identify when a signal falls outside of expected specifications. This can be helpful in identifying outliers and ensuring signal integrity.

Real-World Examples and Case Studies

To better understand the applications of waveform math functions, let's look at some real-world examples and case studies. One common issue in electronic circuits is noise or interference on a signal. By using the FFT function, users can identify the frequency components of the noise and trace their source. In one case, a user was troubleshooting a circuit with an unstable output signal. By using the FFT function, they were able to identify a high-frequency noise component that was causing the instability. Further investigation revealed that the power supply was causing the noise, and the issue was resolved by adding additional filtering components. In another case, a user was troubleshooting an intermittent glitch on a digital signal. By using a math function to differentiate the signal, they were able to identify that the glitch was caused by a fluctuation in the rise time of the signal. By adjusting the triggering settings to trigger on the rise time, the user was able to capture the waveform that caused the issue and quickly resolve the problem.

Best Practices for Using Waveform Math Functions

While waveform math functions can be powerful tools for troubleshooting, it's essential to understand some best practices for using them effectively.

Firstly, it's crucial to understand the function being used and its limitations. Improper use of math functions can lead to erroneous results and misinterpretations. Secondly, it's important to consider the impact of using math functions on the signal's integrity. Applying certain functions, especially ones that involve differentiation or integration, can introduce additional noise and distortions in the waveform. Finally, it's always advised to use multiple math functions to cross-reference results and ensure their accuracy.

In Conclusion

In conclusion, waveform math functions are invaluable tools for analyzing complex signals and troubleshooting issues in electronic circuits. From identifying sources of noise and interference to isolating intermittent glitches, these functions provide valuable insights that cannot be obtained through traditional methods. By understanding the different types of math functions and their applications, along with best practices for using them, users can effectively troubleshoot and resolve issues, ultimately saving time and resources.

Chapter 33: Techniques for Troubleshooting Analog Circuits: Identifying Signal Distortions and Component Failures

Analog circuits are a fundamental part of electronic systems, responsible for processing and amplifying various types of signals. However, as with any type of circuit, problems can arise that can cause signal distortions or even complete component failures. Troubleshooting analog circuits can be a challenging task, as there are many variables and potential failure points to consider. In this chapter, we will explore some techniques and best practices for identifying and troubleshooting signal distortions and component failures in analog circuits.

Understanding Signal Distortions

Signal distortions can occur in analog circuits for a variety of reasons, such as improper component values, faulty connections, or interference from other components. These distortions can affect the quality and integrity of the signal, leading to inaccurate or unreliable measurements. It is essential to understand the different types of distortions that can occur in analog circuits in order to effectively troubleshoot and correct them.

One common type of signal distortion is harmonic distortion, which occurs when the signal contains unwanted frequency components, known as harmonics, in addition to the fundamental frequency. This can be caused by components such as capacitors or transistors with nonlinear characteristics. Another type is intermodulation distortion, which occurs when two different frequencies interact to produce additional frequencies that were not present in the original signal. This can happen when signals from different sources are combined, or when signals pass through components with nonlinear characteristics.

Using Waveform Analysis to Identify Distortions

One of the most effective tools for identifying signal distortions in analog circuits is waveform analysis using an oscilloscope. Oscilloscopes allow you to visualize the shape and characteristics of the signal, making it easier to spot any anomalies or distortions. When troubleshooting signal distortions, it is essential to have a solid understanding of the expected waveform for the particular circuit you are working on. This will help you identify any deviations from the norm and narrow down the potential causes of the distortion.

It is also helpful to use the built-in math functions of the oscilloscope to perform calculations on the waveform. For example, you can use the Fast Fourier Transform (FFT) function to analyze the frequency components of the signal and identify any unwanted harmonics or intermodulation distortions. Additionally, some oscilloscopes have advanced math functions that can isolate specific frequency components or perform signal filtering, making it easier to pinpoint the source of the distortion.

Identifying Component Failures

In addition to signal distortions, another common issue in analog circuits is component failures. Faulty or damaged components can lead to a variety of problems, from excessive noise to complete circuit failure. Troubleshooting component failures in analog circuits can be a tedious process, as it often involves checking individual components and their connections.

One useful technique for identifying component failures is using a logic analyzer in conjunction with an oscilloscope. Logic analyzers allow you to monitor and analyze the digital signals in a circuit, providing insights into the proper functioning of digital components such as logic gates or microcontrollers. By comparing the signals from the logic analyzer with the analog signals from the oscilloscope, you can often pinpoint the source of the problem.

Best Practices for Troubleshooting Analog Circuits

When troubleshooting analog circuits, it is crucial to have a systematic approach and follow best practices to efficiently identify and resolve problems. Here are some tips to keep in mind:

- Start by thoroughly checking all connections and making sure that all components are properly installed and functioning.
- Use the appropriate tools for the job, such as an oscilloscope for waveform analysis and a logic analyzer for digital signals.
- Have a good understanding of the expected signal characteristics for the circuit you are working on.
- Make use of the various measurement and math functions of your tools to analyze the waveform and identify any distortions or abnormalities.
- Document your troubleshooting process and any changes you make to the circuit. This will help you keep track of what steps you have taken and what has been successful in resolving the issue.
- Be patient and systematic. Troubleshooting analog circuits can be a time-consuming process, but being methodical and thorough will lead to more accurate and efficient results.

Conclusion

Troubleshooting signal distortions and component failures in analog circuits can be a challenging task, but with the right approach and tools, these problems can be identified and resolved. By understanding the different types of signal distortions and utilizing waveform analysis and logic analysis, you can isolate the source of the problem and make the necessary corrections. Remember to document your process, and don't be afraid to consult with other experts if you need additional assistance. With persistence and knowledge, you can overcome any issues and ensure the smooth functioning of your analog circuits.

Chapter 34: Advanced Applications of Oscilloscopes

In this chapter, we will delve into some advanced applications of oscilloscopes, showcasing their versatility and wide range of uses. We will explore some case studies and examples of how oscilloscopes are being used in industries such as automotive and medical device testing.

Case Studies in Automotive Testing

The automotive industry is one of the major users of oscilloscopes, particularly in the testing and development of electronic systems. With the increasing use of advanced electronic components in modern vehicles, oscilloscopes have become a crucial tool for engineers in the automotive industry. One popular use of oscilloscopes in automotive testing is in the diagnosis and troubleshooting of engine control systems. The digital signals produced by the engine sensors can be easily observed and analyzed using an oscilloscope, providing valuable insights for engineers and mechanics. Another important application is in the testing of electronic control units (ECUs) in vehicles. Engineers use oscilloscopes to perform real-time testing of the signals and data being sent to various components, such as the fuel injectors and ignition systems. This ensures that the ECUs are functioning properly and helps in identifying any faults or malfunctions.

A recent case study by a leading automotive company found that the use of oscilloscopes in their testing process improved efficiency and reduced their overall production costs. With advanced features such as automated measurements and bus analysis, oscilloscopes are becoming an indispensable tool in the development and testing of modern vehicles.

Examples of Medical Device Testing

The medical industry is another sector where oscilloscopes are extensively used for testing and development. With the growing demand for more sophisticated medical devices, oscilloscopes have become essential in ensuring the accuracy and reliability of these devices. One key application of oscilloscopes in the medical field is in the testing of pacemakers. These life-saving devices require precise monitoring and control of electrical signals, which can be easily achieved with the use of oscilloscopes. Engineers use oscilloscopes to measure the output of the pacemaker and ensure that it is functioning correctly before it is implanted in a patient. In the field of electrocardiography (ECG), oscilloscopes play a vital role in the testing and calibration of devices used for monitoring heart activity. The high-resolution displays and advanced measurement capabilities of oscilloscopes are crucial in accurately measuring the complex waveforms produced by the heart.

A study conducted by a renowned medical research institute found that the use of oscilloscopes in testing medical devices has greatly improved the accuracy of their results. With the ability to capture and analyze high-frequency signals, oscilloscopes have become an indispensable tool in the medical field.

Advanced Triggering Options for Specific Measurement Scenarios

Oscilloscopes have come a long way since their inception, and with advancements in technology, they now offer a wide range of advanced triggering options for specific measurement scenarios. These advanced features allow engineers and technicians to tackle complex and challenging measurement problems with ease and precision. In this chapter, we will explore some of the most commonly used advanced triggering options and how they can be utilized in various measurement scenarios.

Pulse Width Triggering

Pulse width triggering is a powerful tool that allows you to trigger on a specific pulse width of a waveform. This is particularly useful when dealing with digital signals that have varying pulse widths, or when you need to isolate a specific part of a waveform for analysis. With pulse width triggering, you can easily capture and analyze the characteristics of a signal that occurs within a defined pulse width range. To use this feature, simply set the trigger type to "pulse width" and adjust the trigger level and pulse width settings to your desired specifications. The oscilloscope will then only trigger when a pulse of the specified width is detected, making it easier to isolate and analyze specific parts of a waveform.

Video Triggering

Video triggering is a specialized feature that is particularly useful when working with video signals. With this trigger option, the oscilloscope can be set to trigger on specific fields, lines, or frames of a video signal, allowing you to analyze and troubleshoot video signals with ease.

To use this feature, select the "video" trigger type and specify the field, line, or frame you want to trigger on. The oscilloscope will then only trigger when the specified part of the video signal is detected, making it easier to analyze and troubleshoot video signals.

Delay Triggering

Delay triggering is another powerful feature that allows you to trigger on a specific part of a waveform after a specified time delay. This is particularly useful when dealing with signals that have a long pre-trigger time or when you need to analyze a specific part of a waveform that occurs after the initial trigger. To use this feature, simply set the trigger type to "delay" and specify the time delay after which you want the oscilloscope to trigger. This allows you to capture and analyze specific parts of a waveform that may not have been captured with a conventional triggering method.

Runt Triggering

Runt triggering is a specialized feature that is used to trigger on a specific type of signal anomaly known as a "runt pulse." Runt pulses are short, not fully formed pulses that can occur due to signal distortions or interference. With runt triggering, you can easily capture and analyze these anomalies, helping you identify and troubleshoot potential issues with a signal. To use this feature, select the "runt" trigger type and specify the criteria for detecting a runt pulse, such as pulse width or amplitude. The oscilloscope will then only trigger when a runt pulse is detected, making it easier to identify and troubleshoot signal anomalies.

Logic Triggering

Logic triggering is a powerful feature that is particularly useful when working with digital signals. With this feature, you can trigger on specific logic levels or patterns, allowing you to analyze and troubleshoot digital signals with ease. To use this feature, select the "logic" trigger type and specify the desired logic level or pattern that you want to trigger on. The oscilloscope will then only trigger when the specified logic level or pattern is detected, making it easier to analyze and troubleshoot digital signals.

Combining Triggering Options

One of the best things about oscilloscopes is that they allow you to combine multiple triggering options to create a customized trigger. This allows you to tackle even the most complex measurement scenarios with ease and precision. For example, you can combine pulse width and delay triggering to trigger on a specific pulse width after a specified time delay. Or, you can combine video and logic triggering to trigger on a specific line of a video signal when a particular logic level or pattern is detected. The possibilities are endless, and the more familiar you become with these advanced triggering options, the more creative you can be with your trigger setups.

Conclusion

Advanced triggering options have revolutionized the way we use oscilloscopes for measurement and analysis. They provide an additional level of precision and accuracy, making it easier to tackle complex measurement scenarios. In this chapter, we discussed some of the most commonly used advanced triggering options, such as pulse width, video, delay, runt, and logic triggering. As you continue to explore and experiment with these features, you will uncover new and innovative ways to use them in your measurement and troubleshooting processes. So go ahead, explore and unleash the full potential of your oscilloscope!

Chapter 35: Advanced Techniques for Characterizing Signal Integrity

Techniques for characterizing signal integrity, such as using eye diagrams and jitter analysis, are crucial for ensuring the proper functioning of digital circuits. In today's fast-paced and constantly evolving world of technology, signal integrity is a critical component for achieving reliable and high-performance digital designs. In this chapter, we will dive deeper into the advanced techniques for characterizing signal integrity and explore how they can improve the accuracy and efficiency of your measurements.

What is Signal Integrity?

Before we delve into the advanced techniques, let's first understand the concept of signal integrity. In simple terms, signal integrity refers to the ability of an electronic signal to maintain its intended shape and quality as it travels through a circuit. Any disruptions or distortions in the signal can result in errors, loss of data, or even complete failure of the device. With the increasing complexity and higher frequency of digital circuits, ensuring proper signal integrity has become a major challenge for designers and engineers. The signal integrity of a circuit is affected by various factors such as noise, crosstalk, reflections, and impedance mismatch. Hence, proper characterization and analysis of these factors are essential for designing and troubleshooting digital circuits.

Eye Diagrams for Characterizing Signal Integrity

One of the most commonly used techniques for characterizing signal integrity is the use of eye diagrams. An eye diagram is a graphical representation of a series of digital signal waveforms captured over a period of time. It provides a visual representation of the quality and integrity of the signals passing through a circuit. To create an eye diagram, the digital signal is first sampled and then plotted on a graph with horizontal lines representing the bits of the signal and vertical lines representing the sequential samples. As more and more samples are added, the graph begins to resemble an open eye, hence the name "eye diagram." This open eye shape indicates that the signal is well-defined and the circuit is performing as intended.

Eye diagrams are particularly helpful in identifying signal impairments such as jitter, noise, and crosstalk. By examining the eye diagram, engineers can easily pinpoint the source of these impairments and take corrective measures to improve the signal integrity.

Jitter Analysis for Characterizing Signal Integrity

Jitter, defined as variations in the time or frequency of a signal, is a common source of signal integrity problems. It can be caused by a number of factors such as electromagnetic interference (EMI), power supply noise, and thermal fluctuations. Jitter can result in timing errors, noise, or even signal loss, which can significantly affect the performance of a circuit. Jitter analysis is a technique that measures and analyzes the amount of jitter in a system. This analysis helps engineers understand the impact of jitter on signal integrity and identify its source. With advanced oscilloscopes, jitter analysis has become more efficient, accurate, and time-saving. One of the methods for jitter analysis is the eye diagram mask test, which compares the eye diagram with predefined masks and identifies where the signal deviates from the desired shape. This method allows engineers to quickly determine if their circuit meets the required specifications and make necessary adjustments.

Tools for Advanced Signal Integrity Analysis

Apart from eye diagrams and jitter analysis, there are other tools and techniques that can aid in advanced signal integrity analysis. For instance, frequency and time-domain reflectometry (FDR/TDR) can identify impedance mismatches and faults in transmission lines by sending a pulse and measuring the reflections. Vector network analyzers (VNA) can measure the impedance of a transmission line and identify any variations that may affect signal integrity. With advancements in oscilloscope technology, some models now have the capability of performing automated advanced signal integrity analysis. This not only saves time but also eliminates human error, making the analysis more accurate and reliable.

Conclusion

In today's fast-paced world of technology, ensuring proper signal integrity is more crucial than ever. Advanced techniques such as eye diagrams and jitter analysis play a vital role in characterizing and troubleshooting signal integrity issues. With the constant advancements in oscilloscope technology, engineers now have access to more efficient and accurate tools for signal integrity analysis. By utilizing these techniques and tools, engineers can ensure the reliable and high-performance operation of digital circuits.

Chapter 36: Advanced Bus Analysis for Troubleshooting and Debugging

Advanced Bus Analysis Techniques

The use of digital buses in electronics has become increasingly prevalent in modern times. Buses such as I2C, SPI, and CAN are commonly used for interconnecting different electronic components in a system. As technology continues to advance, the complexity of bus systems also increases, making it crucial for engineers and technicians to have in-depth knowledge of bus analysis techniques to troubleshoot and debug issues effectively.

I2C Bus Analysis

I2C (Inter-Integrated Circuit) is a popular synchronous serial communication protocol designed to allow multiple devices to communicate with each other using a minimum number of pins. In recent years, I2C has found widespread use in applications such as computer peripherals, sensors, and embedded systems. However, due to the bus's complex nature, troubleshooting and debugging I2C issues can be challenging. With the advanced bus analysis feature of modern oscilloscopes, engineers and technicians can decode I2C signals and view the transmitted data in a more user-friendly format. This enables them to quickly identify and diagnose any issues with the bus, such as clock stretching or data corruption. Additionally, the ability to trigger on specific I2C events and quickly capture critical bus data helps in troubleshooting intermittent issues, saving time and effort.

SPI Bus Analysis

SPI (Serial Peripheral Interface) is another popular serial communication protocol used to transfer data between microcontrollers and peripheral devices. It is commonly found in applications such as flash memory, sensors, and LCD displays. Troubleshooting SPI bus issues can be challenging, especially with high-speed and complex bus systems.

With advanced oscilloscopes, engineers and technicians can decode SPI signals and analyze the data in a more meaningful way. The oscilloscope's waveform math function allows users to perform calculations on the decoded SPI signals, providing valuable insights into the timing and data integrity. Additionally, advanced triggering and filtering capabilities aid in quickly identifying and resolving issues with the bus, ensuring efficient and accurate debugging.

CAN Bus Analysis

CAN (Controller Area Network) is a popular bus protocol used in automotive and industrial applications for real-time communication between electronic control units (ECUs). With the increasing use of electronic systems in vehicles, the complexity of the CAN bus has also increased, making it challenging to troubleshoot and debug issues. Oscilloscopes with advanced bus analysis features enable engineers and technicians to decode and analyze CAN bus signals accurately. The ability to trigger on specific CAN bus events and filter out unwanted data helps in pinpointing and resolving issues quickly. Additionally, the oscilloscope's ability to visualize the data in both time and frequency domains aids in understanding the bus's behavior and identifying any anomalies.

The Power of Advanced Bus Analysis

The use of advanced bus analysis techniques is crucial for troubleshooting and debugging issues in complex and high-speed bus systems. Without these tools, engineers and technicians would have to manually decode and analyze the data, which would be extremely time-consuming and prone to human error. However, with advanced oscilloscopes, this process is simplified, significantly reducing the time and effort required to resolve issues. Moreover, as technology continues to advance, so do the capabilities of oscilloscopes. Advanced bus analysis features have become a standard in modern oscilloscopes, offering engineers and technicians a powerful tool to analyze, measure, and debug digital bus systems accurately. With the use of these techniques, engineers can design and develop more robust and efficient systems, leading to improved reliability and performance.

Conclusion

In conclusion, advanced bus analysis techniques have become an essential tool for engineers and technicians in troubleshooting and debugging complex bus systems. With the ever-increasing use of digital buses in electronics, the need for accurate and efficient analysis has become paramount. Oscilloscopes equipped with advanced bus analysis features offer a powerful solution to this problem, enabling engineers to quickly resolve issues and ensure the optimal performance of their systems. So the next time you encounter any issues with I2C, SPI, or CAN bus systems, remember the power of advanced bus analysis.

Chapter 37: Advanced Bus Analysis with Waveform Automation

The use of oscilloscopes has become essential in modern-day electronics. As technology advances, the need for more accurate and faster measurements also increases. With the complexity of the circuits and the signals they produce, it can be challenging to efficiently analyze and troubleshoot them. This is where waveform automation comes in.

Using Waveform Automation to Save Time and Increase Efficiency in Repetitive Tasks

Waveform automation refers to the ability of the oscilloscope to automatically perform measurements and analysis on repetitive signals. This feature not only saves time but also increases efficiency by removing the need for manual measurements. With waveform automation, engineers and technicians can now focus on more critical tasks such as analyzing complex signals and troubleshooting issues.

The Benefits of Waveform Automation

Apart from saving time and increasing efficiency, waveform automation offers many other benefits. One of the main advantages is the accuracy it provides. Manual measurements can be prone to human error, whereas automated measurements are more precise and consistent. This is especially important in high-frequency signals where the slightest discrepancy can cause significant issues in the circuit. Another benefit of waveform automation is the reduction of time-consuming tasks. With features such as automated measurements and analysis, engineers can perform tasks that would have taken hours to complete in just a fraction of the time. This not only increases productivity but also allows for more thorough analysis of the signals.

Utilizing Waveform Automation for Advanced Bus Analysis

In the digital world, data is constantly being transferred between various devices through digital buses. It is crucial to understand the behavior of these buses to ensure proper communication between devices. Traditional methods of bus analysis involve setting up numerous measurements and comparisons, which can be time-consuming and prone to errors. This is where waveform automation shines.

With waveform automation, multiple measurement points can be set up within seconds, and comparisons can be made simultaneously. This not only saves time but also provides a comprehensive analysis of the bus behavior. Engineers can quickly identify anomalies and troubleshoot issues, allowing for faster debugging and verification.

Customizing Waveform Automation to Suit Your Needs

One of the great things about waveform automation is its flexibility. It allows users to customize and automate specific tasks according to their needs. For example, engineers can set up automated triggers to capture specific signals and have the oscilloscope perform the necessary measurements and analysis automatically. This eliminates the need for constant monitoring and manual measurements, providing a more hands-off approach.

Not only can tasks be automated, but the results can also be displayed in various formats, such as histograms or tables, making it easier to analyze and interpret the data. This feature is especially useful when dealing with a large number of signals or when trying to identify rare events in the data.

Improving Collaboration and Knowledge Sharing

With waveform automation, engineers and technicians can easily share results and data with their colleagues. Once a set of measurements and analysis has been automated, they can be saved and shared with others. This not only allows for better collaboration but also aids in knowledge sharing. Engineers can now build upon each other's work and improve efficiency in the long run.

In Conclusion

In today's fast-paced electronics industry, time and efficiency are essential. With the ever-increasing complexity of electronic circuits, it can be challenging to keep up with measurements and analysis manually. Waveform automation offers a solution to this problem, providing accurate and efficient measurements and analysis. By utilizing waveform automation, engineers and technicians can not only save time and increase efficiency but also improve collaboration and knowledge sharing within their teams. So why not give it a try and see the benefits for yourself?

Chapter 38: Power Integrity Measurements

Overview of Power Integrity Tests

Power integrity is a critical aspect in the design and functioning of electronic devices, as it ensures that the power delivery networks are capable of providing stable and reliable power to the components. Power integrity tests involve measuring and analyzing various parameters such as the voltage, current, and power consumption of a device, to ensure that it is operating within its desired specifications. In this chapter, we will discuss the different power integrity tests and how to perform them using an oscilloscope.

Power Rail Measurement

Power rail measurement is the most common type of power integrity test, as it involves measuring the voltage on the power supply rail to ensure that it is within the desired range. This test is crucial in identifying issues such as voltage drops, noise, and ripple on the power rail, which can affect the performance of the device. An oscilloscope is the ideal tool for power rail measurements, as it can accurately capture and display the voltage waveform in real-time.

Power Consumption Analysis

In addition to measuring the voltage on the power rail, it is also essential to analyze the power consumption of a device to ensure that it is within the expected range. Power consumption analysis involves measuring the current draw of a device and calculating the power consumption using Ohm's law. An oscilloscope, with its ability to measure current and voltage simultaneously, is an excellent tool for power consumption analysis.

Performing Power Integrity Tests with an Oscilloscope

To perform power integrity tests with an oscilloscope, one must first ensure that the instrument is properly set up. This includes selecting the correct voltage and current probes, as well as ensuring that the oscilloscope is properly calibrated. The next step is to connect the power supply and the device under test to the oscilloscope using the appropriate probes.

Once the setup is complete, the oscilloscope can be used to capture and analyze the voltage and current waveforms. The various triggering and analysis features of the oscilloscope, such as pulse width and rise time measurements, can be utilized to identify issues such as noise, ripple, and voltage dips on the power rail. Additionally, the math functions of the oscilloscope can be used to perform calculations such as power consumption, making it a versatile and powerful tool for power integrity tests.

Best Practices for Power Integrity Measurements

To ensure accurate and reliable power integrity measurements, it is essential to follow best practices while using an oscilloscope. These include using high-quality probes and connectors, ensuring proper grounding of the oscilloscope and the test setup, and paying attention to signal integrity. It is also crucial to use the appropriate settings and functions of the oscilloscope and to properly interpret and analyze the obtained measurements.

Conclusion

As the complexity of electronic devices continues to increase, ensuring proper power integrity has become more critical than ever. Power integrity tests, such as power rail measurement and power consumption analysis, are vital in identifying and resolving issues that can affect the performance and reliability of a device. Using an oscilloscope to perform these tests not only provides accurate measurements but also allows for in-depth analysis and troubleshooting. By following best practices and utilizing the full capabilities of an oscilloscope, one can ensure that a device has proper power integrity and can function as intended.

Made in the USA
Las Vegas, NV
25 November 2024